THE COEXISTENCE TRILOGY – PART ONE

FOREVER

FOREVER

A LEGAL SCI-FI STORY

DANIEL GERVAIS

ANTHEM PRESS

Anthem Press
An imprint of Wimbledon Publishing Company
www.anthempress.com

This edition first published in UK and USA 2023
by ANTHEM PRESS
75–76 Blackfriars Road, London SE1 8HA, UK
or PO Box 9779, London SW19 7ZG, UK
and
244 Madison Ave #116, New York, NY 10016, USA

© Daniel Gervais 2023

The author asserts the moral right to be identified as the author of this work.

British Library Cataloguing-in-Publication Data
A catalogue record for this book is available from the British Library.

Library of Congress Cataloging-in-Publication Data
A catalog record for this book has been requested.
2023934296

ISBN-13: 9781839989117 (Hbk)
ISBN-10: 1839989114 (Hbk)

ISBN-13: 9781839989124 (Pbk)
ISBN-10: 1839989122 (Pbk)

This title is also available as an e-book.

To my girls

There shall be no more death, neither sorrow, nor crying, neither shall there be any more pain: for the former things are passed away.

(Revelation, 21:4)

All of us, all of us will perish

(Sergei Yesenin)

PREFACE

This book uses (science) fiction to accompany the reader in thinking about what will happen when nonhuman entities able to perform many of the tasks associated with human higher mental faculties—the very same ones that have until now put humans at the pinnacle of the species on Earth and allowed humans to use, transform and misuse other species—start living among us.

Fiction is particularly useful in legal education because, as John Wigmore noted, it provides a *"catalogue of life's characters*. And the lawyer must know human nature. He must deal with its types, its motives. There he cannot find—all of them—close around him; life is not enough."* Fiction can both allow a student to identify with a "real-life" situation and allow them to understand a situation they have not and may well never encounter. Fiction can also inform how one will react faced with a situation never before encountered in real life but experienced vicariously through a work of fiction.

Using fiction is helpful more specifically in a legal education context because laws as they stand in any place and time reflect the socioeconomic, cultural, geographic, and political contexts in which those laws were created. The law seen globally as a system of rules meant to regulate behavior will change, therefore, as those contexts evolve, either to follow or resist change. Moreover, embedded in laws and legal systems is a system of values. Up until now, humans alone have been making all the decisions about those values.

Among the various genres of fiction that can be used as a pedagogical tool, science-fiction offers distinct advantages. Science fiction might be described as the nonfiction of tomorrow, and this explains its appeal in law schools. For example, Fuller's *The Case of the Speluncean Explorers*† has long been used to

---

*John H. Wigmore, 'A List of One Hundred Legal Novels', (1922–1923) 17 *Illinois Law Review* 26, at 31.
† Lon L. Fuller, 'The Case of the Speluncean Explorers', (1949) *Harvard Law Review* 62.

discuss natural law and positivism and the relationship between law and morality. Science-fiction is situated in a world that does not (yet) exist, but it often has deep links to some of the most disturbing aspects of our society. W.E.B. DuBois's *The Comet*, for example, is a story of racism in America and Derrick Bell's *Space Traders* discusses "not only the United States' history of slavery and Jim Crow, but also Indigenous removal and the incarceration of Japanese Americans during WWII."‡

Science fiction opens an eccentric—that is, away from our temporal center—window on human history yet to be written that stands in a house built on our collective past. It contains echoes of widely divergent outlooks, often the result of small choices about the individual paths we follow, often not even consciously. Science fiction ponders the myriad paths that our collective future can carve in the sands of time, only to be washed away by the tides of years passing or captured as distorted images on the mutable pages of history books. Science-fiction is the scout of possible futures often hoping to chart which ones may constitute progress, because change is inevitable, but progress is not.

Well beyond legal education, anyone interested in ethics, the progress and philosophy of science, cultural changes, social justice, political science, and many other areas can use science-fiction to reflect on whether the current path of technological progress is, in fact, a form of *human* progress.

Then this book weaves in poetry, including poetry embedded in film. Poetry is a form of human expression that AI machines can produce, but they are very far from the level of great human poets. Will they ever reach it? Poetry is in a way science fiction's travel companion in this story. It opens a window on the human soul: past, present, and future. As we constantly reconstruct narratives to explain our past, individual and collective, poetry stands vigil. It suffuses words with layers of meaning unknown to the reader who discovers them, until they do. It shines its dark light in the deepest recesses of the self and helps us see more vividly into the heart of the human enterprise. Devoid of the shadows of unnatural order that the pallid light of reason shines on the affairs of the human soul, poetry pierces the veil that conceals that variegated admixture of science and belief we call truth. Poetry's sinewed path through time and space tears the thin, rational fabric with which history books are woven, and too often undone, as if written on Penelope's loom. No wonder many great poets could

---

‡ Ellie Campbell and Antonia Eliason, 'Teaching Law and Science Fiction at the University of Mississippi', (2022). 52:1 *SFRA Review* 165, at 166.

not stand to live with other humans after digging themselves out of Plato's cave, one verse at a time, and took their own lives instead.

Combining poetry and science-fiction makes sense when they are seen as the tools that humans have used for thousands of years to understand their past and predict their future. We can predict the future—up to a point—but can we choose the way forward? Isn't life, for each one of us, but a haphazard path, and human history merely the aggregate result? Science fiction and poetry remind us that we exist because we search for our true selves. Not because the object of our search necessarily exists, but because the subject does.

Teachers who want to explore the book with their students and interested readers will find a list of questions (e.g., for discussion, self-reflection, or student essays) at the end of the book. Links to material (songs, poems, pictures) mentioned in the book are available at *danielgervais.net*.

ACKNOWLEDGMENTS

Writing fiction is a lone endeavor but then there are many people without whom this project would not have come to fruition. They are the loved ones, friends, and colleagues who encouraged me, provided tons of helpful suggestions and made this book better, such as it is. Svetlana's continuous encouragement and support were crucial to get this book to the finish line. I cannot thank her enough. To my daughters Daphne and Sophie, thanks for always believing your dad could do this.

A rough first draft of this book was read by many of my colleagues at Vanderbilt, the informal Sci-Fi Book Club, with professors from several parts of campus. I am particularly grateful to Paul Edelman, Carolyn Floyd, Rob Mikos, Robert Scherrer, and Yesha Yadav.

I will be forever (!) grateful to Alissa McGowan who helped me edit and improve the book and provided myriad pieces of advice along the way.

I was lucky to be able to rely on my former student Jin Yoshikawa to discuss the use of "san" and other Japanese words (*kun, sensei* etc.). In the end, I decided *not* to follow Japanese practice to the letter, as would almost certainly happen if Americans were to decide to use "san" to replace gendered pronouns.

I am similarly grateful to Beibei Yang, a former student of mine, who checked my choice of Chinese terms. I'm so glad she did, but any mistake is entirely mine.

Finally, it was reassuring to read that my crazy idea for the book seemed no longer that crazy when I came across Wendell Wallach's idea of "mind files." His work on the intersection of law and AI has been and continues to be an inspiration.

For the cover art, my thanks go to the very talented Anna Messer.

Except for well-known quotes, translations of Russian poems were initially made using DeepEl and edited by Svetlana Yakovleva.

PART I

2037

CHAPTER 1

She woke up sweaty and a little confused, thinking about death. Again. The digital clock next to her bed read 4:01 a.m.

Was it another wave of grief pounding the barren shores of her asphyxiated soul? Christine's mother had passed away two years earlier after what the media likes to call a "battle" with cancer. It may have been a battle, but her mother wasn't the soldier, she was the battlefield. Christine had been through wave after wave of grief, each one like a semiopaque tunnel. She had kept walking, and days seemed brighter now. Some days she didn't even think about her mom. But most days, she did, and fell back into a deeper, darker part of the tunnel.

Or was it the futility of existence? Spending her life filling her brain, from school to university to postdoc to her current job as a law professor, by reading hundreds of books and countless journal articles. Once you're dead, the maggots won't know the difference between the brain cells that can parse Plato's teachings and those that are only interested in the latest celebrity scandal du jour. Why bother? And yet, Christine spent hours at the gym delaying the inevitable.

Was her mom happier where she was? Was she anywhere?

Dewey, who had been sleeping at her feet, was now fully awake. He put his fur in her face and started purring, sketching a faint smile on Christine's face. She wished there was someone she could talk to. It was way too early to call her best friend Rachel, even though Nashville was an hour ahead. She briefly thought about calling her dad in Huntsville, but they hadn't spoken in so long. She swallowed hard at the thought and was overcome by another upsurge of sorrow.

Turning toward her night table, with its scuffed olive-green paint to make it look vintage aughts, she opened the drawer and popped a small white pill. She reached for the glass of water she put on the table every night before going to bed and swallowed just enough of the tepid liquid to push the pill down.

***

"Good morning, Christine!" The feigned cheerful voice of her home robot, Harry, pulled her abruptly from sleep just a few hours later. "It is going to be a beautiful fall day today. Mostly sunny and a maximum of sixty-four degrees. Time for breakfast!"

Still groggy from her 4 a.m. pill, Christine opened her eyes to find Dewey looking right at her. He always waited for her to get out of bed, meowing for food as soon as Christine was up. She moved the heavy duvet and tried to align her feet with her fluffy cat slippers, then made her way to the adjacent, white-tiled bathroom. After splashing cold water on her face a few times to gather her courage, she looked at herself in the wall-sized mirror. Her reflection screamed for concealer to hide those bags under her eyes and even out her pasty white skin.

It only took her fifteen minutes to get ready. She never wore much makeup and kept her red hair very short. She liked that she didn't need to blow dry it. Freshly showered and with her hair towel dry, Christine sat down at the breakfast table, where eggs, toast, and coffee were waiting for her.

Putting her watch close to her mouth, she said, "Maya, news." A screen on the wall turned on, and the voice on her watch said, "Christine, your weight this morning is 137.5. That is 1.5 pounds above your target. Enjoy your breakfast." She hated that feature but put up with it because turning it off would increase her life insurance costs.

Images of a city in ruins appeared on the big screen. The robot reporter's voice said, "President Lopez had no choice but to order the destruction of part of the city of Tehran after the launch from Lebanon of a drone and rocket attack on Haifa at ten-thirty p.m. local time last night. The attack in Haifa destroyed 720 homes and killed 4,000 Israelis, many of them Arabs. The US attack in response was carried out by drones and missiles launched from our base in the Free Republic of Turan."

The United States and her allies in the Gulf had been able to foment a coup a few years earlier and split Iran in two parts: The Northern Islamic Republic of Iran, with Tehran as its capital, and the Free Republic of Turan, with its capital Shiraz. Turan was aligned politically with the United States and had agreed that a US military base could be set up there.

Christine listened with half an ear, but she had become so desensitized to the news. It was nothing but wars, planetary catastrophes, corruption, waning democracies everywhere including in her own country, and politicians seemingly both unable and unwilling to do anything about it, like driving a car into a wall and hitting the gas pedal instead of the brakes. She had little faith in her fellow humans.

Leaving the breakfast dishes for Harry to clean up, she went back to the bedroom to get dressed. She had a thing about matching at least one piece of clothing with her hair and her blue-green eyes. Today she picked a pair of black pants purchased in Italy, with thin stripes that matched her hair.

When she finished getting ready, she called for a PC to the law school.

CHAPTER 2

As the PeopleCar was stopping in front of Christine's townhouse, her first—and only—true love, Paul Gantt, was arriving with the first rays of morning sun at Eidyia headquarters outside of Portland, Oregon.

The weather was unusually warm for September, and, in obligatory tech normcore fashion, Paul was wearing tight black jeans and an old, but ridiculously expensive, t-shirt. Once the sensors identified his face and read his S-Chip, the main door opened, and he entered the building and proceeded down a long hallway. As cofounder and Chief Transhuman Technologist at Eidyia, a company he had named after a Greek goddess of knowledge, Paul had been the one to develop and deploy the S-Chips now inserted in a vast majority of Americans.

Paul arrived in front of a large metal door near the center of the building. Like the walls, it was made of a two-inch-thick steel alloy to prevent eavesdropping by drone, satellite, or otherwise. It also prevented any S-Chip data from being collected while inside that space.

The door opened automatically, and Paul walked into the small entry-lock area. Once the door behind him had closed and the system had detected only one person, the second door opened. He was in. Across the room, Eidyia's cofounder Bart van Dick was looking at a wall of screens. He turned, raising an eyebrow at Paul's somewhat disheveled hair.

"Hi, Paul! Late night?" Bart looked as cool as usual, of course. Paul attributed his friend's even keel to his Dutch upbringing, learning to live by constantly windy shores and stormy waters.

"Yeah. I was reviewing the latest dataset on sensory perception for the new skin. It's very good. The nano-receptors are working better than expected. Those nanoparticles are getting really good at fixing tears in the skin! I also have good news on the sexual receptors. I think they will work as planned. With that, we can probably finish the first complete model in a few weeks."

What Paul did not mention was that he had spent the night in the dark and stormy clouds of depression that often enveloped his mind. He had increased

4

his antidepressant dose beyond the maximum printed on the bottle and, with a massive dose of caffeine added to the mix, felt almost ready to face the day.

"Jeremy is still working on kinks in the inference engine for personality data transfers, but that is also more or less on schedule," Bart reported. "If all goes well, we will be able to go public with an announcement of the first full model early next year!"

"That's great," Paul said, wishing he could feel any real enthusiasm about the project that was his life's work. "I'll get back to work on the new generation of synthetic skin. Let's catch up later."

"Sounds good," Bart replied, his eyes already glued back to the screens in front of him.

CHAPTER 3

Christine was a bit of a celebrity at the law school. Her casebook, *Robots, AI & the Law*, was now used at over 50 law schools across the nation. She enjoyed writing, but the full measure of her mental energy was deployed in the classroom, and a law school classroom was a special experience—and a demanding one. A seasoned litigator had cried in her office after teaching a guest class a few years before. "Harder than any appellate court I've ever seen," he'd said. And it could be. A good class kept everyone, student, and instructor, on their toes, minds connecting.

She thought of classes as baseball games, she and the students pitching questions at each other. But she had to steer in a way that they both could win, the students having progressed in their thinking about the law, and her going back to her office and a well-deserved cup of tea with that feeling of self-fulfillment that a good class could supply, the satisfaction of a mission accomplished.

She started her class that morning as she started every class, with an easy open throw to the entire group.

"When is a robot liable for its actions?"

None of the twenty-five students in front of her caught the ball.

Her eyes roamed the classroom, calculating who she could call on to get the game going. Her gaze settled on a petite Chinese American woman sitting at the end of the first row who looked younger than the average twenty-four-year-old law student.

"Weijia-san, when is a robot liable?"

"That depends. What do you mean by 'robot' and then 'liable' for what?"

Good, Christine thought, reminded once again that many of these students were already well-trained as future lawyers. Answer a question with a question. "Weijia-san, you can't define robot? We've talked about it many times. It should be easy by now."

She mentally thanked her law school for officially switching to the Japanese-inspired *san* a year before to refer to all human beings, instead of having to find her way through the gendered soup of the 2020s. Students were still required to use *professor*, although some had started using *sensei*, which no one opposed.

Weijia looked down for a second, as if looking for words under her desk. She perked her head up, smiling. "One accepted legal definition is—Weijia looked at her notes—that 'a robot is an AI system with agency, capable of learning and making decisions based on knowledge it was given or has acquired, and embodied in physical form, often with anthropomorphic quality.'"

She was quoting the UN definition from the casebook, and she was right about "agency," the autonomy of robots. That was all the law seemed to care about, whether machines "acted" like humans. Psychological research had shown that physically embodied AI systems, such as Eidyia robots, interacted much better with humans when they had human form. Basically, people trust robots that have two legs, two arms, and two eyes. Flip this, and that explained why people distrust machines that looked like, well, machines.

"Very good. Thank you, Weijia-san." Christine took a few steps to the left and faced the other half of the class. "So now that we know what robots are, what are they liable for? Let's make it more concrete. If a robot hits me by mistake and breaks my arm, what does the caselaw say, Andre-san?"

Andre Prudhomme was shy, but Christine knew them to be highly intelligent and detail oriented. Christine suspected they wore those trendy glasses, designer jeans, and gender-neutral shirts to look more au courant than their upbringing in rural Louisiana might suggest to their classmates. True to form, they replied, "We know that the robot manufacturers are not liable under most state statutes, as they have been exempted by state laws going back decades to the first Nevada state statute on autonomous vehicles."

"Good start, Andre-san. But my question was about cases." Her attention was drawn to Nadia Patterson, fidgeting on her seat in the third row. Nadia's usual conservative wardrobe of flowy skirt and pastel-colored, loose-fitting blouse fluttered with her movements as she passed a paper bag to her neighbor.

"Nadia-san, can we help?" Christine asked, annoyed but trying to remain Zen.

"Sorry, Professor. I was just passing a bag of sweets around to the other students. My mom sent them to me. They're from Lebanon."

"Sweets?"

"Nougat with rosewater, cardamom, and nuts."

Christine's face softened in an instant. The thought of nougat brought to mind vivid images of a trip to Istanbul with Paul—walking in the Grand Bazaar with its ancient arched ceiling, a million novel smells, and learning from a shopkeeper selling all manner of Turkish sweets the differences between nougat, lokum, and malban.

"Hmm, if you happen to have an extra one …"

"Of course." Nadia handed Christine the bag.

Christine plucked one piece out and sank her teeth into the soft, small square of nougat topped with pistachios and walnuts. There was something about the texture of fresh nougat that was both unusual and comforting. And then cardamom, a taste at once familiar and new.

After each student had had a chance to pick from Nadia's bag, Christine turned back to Andre.

They grinned. "I was just getting to the cases, Professor. But I must say, this is really good."

"I agree. Thanks so much, Nadia-san." Another thought crossed Christine's mind. How many calories did she just ingest? She made a mental note to add a few minutes to her next run.

"So, there are three groups of cases on robot liability," Andre began. "The first group, exemplified by *Jones v. Strasburgh*, a 2029 Supreme Court of Massachusetts case, treats advanced robots like animals and imposes liability on the owner if the damage was caused directly or indirectly by the owner's instructions or programming. The second group of cases, as in *Kanes v. Simmons* at the Tennessee Supreme Court, applies the rules of guardianship, treating robots as children, but the practical outcome is essentially the same as the first group. The third group includes *Robertson v. Chadwick* and *Obrador v. James*, two cases of the California Supreme Court that have held that robot owners are not liable unless the owner or operator has specifically and directly instructed the robot to perform the action that generated the harm."

Straight from the casebook. "That is an extremely detailed answer, Andre-san. Let me follow up to make sure everyone's on the same page. The California cases reflect the Safety Algorithm Standards adopted by robot manufacturers to prevent robots from acting in a way that causes harm to humans, correct? What do you think of the SAS?" She rested her gaze on Andre once again.

"SAS, or 'SAS1' as most people call it now, is a very basic document, kind of like Asimov's old laws of robotics. Those did not work, of course."

"Oh? Remind us why, please," Christine asked, sitting on the corner of the old wooden table near the front of the room, worn by years of Socratic teaching.

Andre pushed their blue-rimmed glasses up their nose with one finger. "There are many reasons. First, robots are now routinely used in law enforcement and by the military. The first Asimov Law, which went something like 'a robot may not injure a human being,' can be thrown out the window, because police and soldier robots often cause harm, but in principle for a good cause."

Christine looked around the room. Nadia was shaking her head, her long, blue-black hair moving gently with the motion. "Nadia-san, do you disagree with Andre? Or is it the nougat?" She smiled at Nadia, who always seemed to be doing three things at a time, even when she wasn't moving. "Did you want to say something in response to Andre?"

"It's just that I was thinking of a discussion I had in one of my undergrad ethics classes." She grabbed her bag and pulled out another piece of nougat, then put it back and looked at the built-in tablet in front of her. She didn't need to log in, as the tablet recognized her from her S-chip.

"What was that discussion. Nadia-san?" Christine prompted.

"I'm trying to remember it. I think the professor asked us to imagine that there was, like, a system capable of predicting crimes or aggressions, and that it would have the ability to stop and, if necessary, like, kill anyone about to commit a crime. That would, like, put an end to crime, which everyone would agree is a plus."

Christine knew this example well. It reminded her of an old movie, *Minority Report*. It was also a classic example whenever Asimov's Laws of Robotics came up. "But that implies a direct violation of Asimov's first law, doesn't it? His second law was that robots must obey humans, except if it violates the first law."

Nadia moved in her chair again, and Christine sensed they were going somewhere now.

"Wasn't Asimov assuming that people would be benevolent when giving orders to robots?"

Crickets.

Nadia again grabbed a nougat from her paper bag and this time took a minuscule bite. When she had almost finished chewing it, she looked at Christine again. "I'm not sure. If any form of physical or economic injury is, like, covered under the first law, this would mean that a robot would not obey an order that creates that kind of harm. Maybe that would prevent robots from causing any harm, but then, another problem with the second law is whether the robot, like, *knows* it is going to cause injury."

"So, what do you think of the draft SAS2 standards?" Christine asked. "Can you compare that to Asimov's Laws?"

"I think SAS2 is, like, far better," Nadia said. "It limits the obligation not to cause harm that is neither inevitable nor necessary. A police robot could cause some harm if that is necessary to prevent a crime, for example. But then we would be applying to robots a much higher standard than to humans. Besides, most robot owners now have, like, robot insurance, and in most cases directly from Eidyia. And Eidyia's lawyers rarely lose! I think everyone in this room would love to work for them!"

"Not me!" someone piped up from the back row.

Christine knew before looking up that the voice belonged to Mira, her most vocally anti-robot student.

Mira pushed her pink hair back with her right hand, perhaps in anticipation of casting off any counterargument. "Eidyia is a frigging monstrosity. Probably almost everyone here is wearing their spy chip!"

"Oh, c'mon Mira," Mary interrupted, "that chip is keeping people healthy."

Christine stifled a sigh. *Here we go again.* Christine liked that Mira added sizzle to class discussions, but sizzle can quickly turn to burn.

"Maybe, but at what cost?" Mira fired back. "Do you even have a life of your own, or do you just do what the chip tells you to do?"

Christine had put an end to too many heated exchanges between these two to count and knew it could easily poison the entire class if she let it continue. She stood and moved closer to the students, as if to affirm her authority. "That is not a debate for today. What do you think of SAS2, Mary-san?"

"Well," the sporty blonde said, looking at Mira, "I wish *people* functioned according to those rules! According to SAS2, a robot must both courteous and effective in communications with humans. No rudeness, no feelings hurt, period. That is the way it should be."

"Yeah, right," Mira quipped. "No debate, no discussion, just the mirror of your own thoughts. What progress!"

"Mira-san," Christine said sharply, "please let Mary finish. Mary-san, let's get back to the liability issue. Do you think robots *should* be liable for their actions?" She had pronounced "should" slowly for emphasis. Mary's hands were shaking a bit and her cheeks were flushed. Christine gave her a sympathetic look. "Okay, let's come back to you later." Her eyes scanned the rest of the class. "Jerry-san, what do you think?"

Christine rarely called on Jerry Silverston. She knew he tried to hide in the back row, and she had more empathy since he had confided in her that he was in law school because his father had put pressure on him, but he hated every minute of it. He even dressed *not* to fit in, wearing cheap jeans two sizes too big and old t-shirts. He'd told her he had taken this class because it seemed less boring than the others but had admitted to Christine that he often had trouble following the discussions. All he knew was that robots had eliminated most of the chores he had to do growing up, and his AI-powered computer could easily find case summaries so he didn't have to read those interminable court decisions. Christine felt a soupcon of regret each time she called on him, but she wanted him trained as a lawyer, after all, and lawyers needed to be able to speak in public.

"Ah, hmmm. Well, that depends," Jerry managed to utter, his face turning red. Christine wasn't worried. This happened each time she called on him. She was hoping it would diminish with time.

"That might look like a safe answer, Jerry-san," Christine said with a reassuring smile, "but it's a non-answer, unless you can tell us what it depends on."

Jerry visibly relaxed a bit. "Well, robots are not human, so they cannot be liable like us."

"Okay, but can they be liable *not* 'like us'?"

"Well, they are better than us in so many ways." Jerry glanced at Mira, sitting next to him rolling her eyes. "They make mistakes, I guess, but never intentionally."

"Ah, so you think robots have intentions?" Christine asked. "Isn't that reserved for human beings?"

"Well, yes. I mean, I didn't mean, like, intention intention," Jerry muttered. "I meant, like, they don't intend to do harm."

"Sounds like intention to me, Jerry-san. Who wants to help out on this one?" Christine went back to half-sit on the corner of the small table.

"I think what Jerry is trying to say," Charles said from the other side of the room, "is that robots do not have a conscience, so they cannot distinguish good from bad and cannot form the intent to cause harm deliberately."

Tall and a little heavy, Charles Brassel wore his hair very short and kept a precisely maintained three-day beard. He had won the Black Law Student Society's Best Student Award last year, and Christine thought of him as one of the smartest persons in the room. But sometimes he needed to be pushed a bit.

"Good and bad, hmm?" Christine said. "What do you mean by that?"

"Well, humans can develop their own set of morals and make decisions accordingly," Charles explained. "They can then be held liable for their decisions."

"Not so," a voice cut in, and the class's focus shifted to Esther, a plain young woman with pale skin, medium-length brown hair, a mole on her right cheek, and heavy, dark-rimmed retro glasses. "We know from behavioral research that people make two types of decisions. Some are made without thinking, like when you try to keep balance after tripping on something. Then there are decisions that are more deliberate. The law says we can be held liable for both."

"I don't disagree, Esther-san," Charles said. It is true that humans and robots don't decide the same way, or that humans have more than one way to make decisions."

"Interesting discussion," said Christine. "That reminds me of the famous trolley problem."

"The trolley problem?" a student asked from the back. Jorge Sanchez, wearing his trademark long-sleeved shirt and khakis.

Christine knew just who to turn to. "Nadia-san, what is the trolley problem?" Christine had seen Nadia's CV and knew that with a BA in ethics and social responsibility, she would have the answer to this one.

Nadia folded her arms, relaxed and confident as she answered. "There are, like, multiple variations of this problem, but essentially, it's like a classic in the literature about ethical dilemmas. Assume that a trolley is going downhill and, like, the brakes stop working. The trolley conductor has two choices: turn right or left. If the trolley goes right, it will certainly kill one person. If it goes left, it will, like, hit and possibly kill up to five. In some variations, the single person is a child and the group of five is composed of, like, older adults. We can also vary by, like, gender and stuff, for example. So, the question is, which option is better as an ethical matter?"

Christine nodded. "So, Nadia-san, tell us, what would a robot do?"

"Initially, programmers tried to avoid letting them choose. Owners of self-driving vehicles had to answer a series of ethical questions, and those choices were programmed into the car. Now, statutes provide that owners of robots programmed according to SAS almost never have any liability."

"That's correct, but also non-responsive. My question was, what would a robot do?"

Nadia unfolded her arms and furrowed her brow. "Hmm, like, I must admit, I'm not sure."

"I think that's actually easy," Mira piped in loudly. "It's a robot. It thinks like a calculator. It would just multiply the probability of harm, the level of harm, and the number of people."

"Are you sure?" Christine asked, smiling to try to lower the tone, but not wanting to stop the discussion.

"Actually, if I may."

Christine turned to Roger Farha, the class's self-appointed resident expert. He mentioned his MIT degree in AI and robotics engineering more often than necessary, and Christine didn't like his know-it-all attitude, but having someone she could call on like an online dictionary was useful at times. He was in law school because he'd been offered a full ride and thought he would try it, to see if the law was good enough for him, but then actually started liking it after Christine had convinced him he could become a patent lawyer. Although now, she wondered whether a call to the Bar was in Roger's, or society's, best interests. She took a step toward him, her mind already pouring water on a nice tea bag back in her office. "Go ahead, Roger-san."

"Robots are able to learn what people consider good and bad because our reactions to other people's actions tell them what we think is good or bad. So, they learn from us, not so much as individuals, but collectively. In the trolley example, I'm not sure that either option is 'good,' so you're picking between bad and bad."

"Maybe the point is to avoid the situation in the first place," a muscular blond man seated in the third row suggested.

Tommy Neuendorfer came from a wealthy New England family. Although he was usually wearing Brooks Brothers, he'd also been known to show up wearing flip flops and an old t-shirt—but always sported perfectly coiffed hair. He was clearly popular, knew he was good-looking, and seemed to do well at everything effortlessly. As far as Christine was concerned, he had privilege written in neon lights on his forehead. Usually for her that created distance. She was in favor of merit-based admissions, but also knew that some students had to work harder to have their merit recognized. When she'd learned that Tommy was spending two long nights a week working in a clinic that provided free legal services to victims of domestic abuse and immigrant workers, where he always went the extra mile, she'd changed her mind about him. It may have been CV-padding on his part—maybe even preparation for a future political career—but there were easier ways to do that.

Christine nodded and gestured for him to continue as a student in the second row looked past her, probably at the big digital clock on the wall. *They're getting tired*, she thought. *Me too.*

"Look, with the hundred-percent tax on personal cars," Tommy said, "we're down to I think less than nine percent of people still driving their own vehicle, and then almost all those cars except those old pre-2029 models have a full self-driving mode. We are down to less than a few hundred major accidents per year—that's, like, a ninety-five percent reduction from the days when people used to drive themselves. And as we find ways to make more car sensors weather-proof, we'll probably be down to less than a hundred within a few years."

"All those numbers sound about right to me, Tommy," Christine said. She peeked at the clock. One minute left. Time to wrap up. "But that means that a lot of people are still getting hurt or worse, and that in some cases, self-driving cars *have to* choose. We still need rules for those cases. On that note, see you next week. Don't forget to turn in your written assignment on SAS 2 by Friday."

Christine stood and waited for the students to finish packing up their things and filing out before heading to her office for that cup of tea.

CHAPTER 4

On Wednesday, it rained all day in Oregon. At Eidyia HQ, Paul was looking at several graphs on the three screens in front of him, touching a piece of the new artificial skin and reviewing readings on the screen. He then compared them to readings taken from his own arm. They were close. Millions of nanoparticles of various types embedded in the artificial skin could feel temperature, touch, and wetness faster but almost in the same way a human could.

Paul could already picture himself in a new body. Many people had accused Eidyia of using the term "transhuman" abusively, saying the S-Chip did not make its host a "cyborg." What critics didn't know was that the S-Chip was a means to an end. *True* transhumanism was Paul's real mission. He wanted the ability to replace people with a copy that would never go hungry, never get sick or tired. A body that would never need sleep and could be repaired when necessary. And when that body deteriorated, the "person" could be transferred to another new body. Forever. The plan was for the new body to look just like the person it was "replacing."

Paul had never discussed in detail with Bart—or the others who, like him, were devoting almost all their waking hours to creating human robots—what would happen when they succeeded.

Paul made his way back to the small kitchen and pressed the button to get the Simonelli ready for action. The machine was hard at work producing a perfect cup of espresso when Bart walked in.

"I am so glad we got this coffee machine," Paul said. "That old one was crap."

"Well, expensive crap, but yeah, this one is so much better." Bart sat down at the small table reserved for the few employees allowed in this part of the lab. "Can you make me one?"

"Sure thing."

When the Simonelli finished brewing, Paul brought both mugs over to the table and sat down.

Bart smelled the coffee in his cup, a look of nostalgia softening his face. "Remember that cheap old Bialetti stovetop we used when we lived at your parents' house? That thing made coffee as good as any of these fancy machines, if you ask me." He took a sip and frowned. "Yeah, definitely."

"I'm not sure I agree." Paul never liked to let his mind wander back to the old house with the gambrel roof and wood siding that had been white in a previous life where he grew up in Lowell, Massachusetts. Whenever he thought of that house, he remembered the horrible old crone who lived in the hovel next door. She always seemed to be spying on his brother and him whenever they were playing in the small yard or on the street, disapproving of youth every foul breath. Paul deeply hated those memories—himself as a gamin doing wheelies around the working-class neighborhood on his yellow bike with its glittering banana seat, his alcoholic father and depressed mother, and the day he'd borrowed John's motorcycle and crashed it into a tree. He would never admit it, but he suspected it was damage to the front wheel that had caused the high-speed crash on I-90 that had killed his older brother.

Bart tilted his head towards his cup. "Going from broke to billions of dollars means you pay more for everything, not that everything is better." He paused and took another sip, as if to confirm that the coffee wasn't worth the money. "So, how's the new batch of skin?"

Paul was happy to get back to business. He really didn't want to think about the past. "I think we've got it just about right. I'm almost ready to sign off on the live trial. It's that good."

Bart grinned broadly and slapped him on the back. "Wow, Paul. Great job! Those nanoparticles are doing the job then?"

"Yes. In fact, they are more precise and powerful than human skin, but I found a way to dumb them down just a bit! Other than the fact that it won't be able to form goosebumps, the new skin is damn close to perfect. Much more resistant to heat and tear, too. The only thing that concerns me is that it looks a bit different when wet. It's kind of shiny."

"Well, who will notice?" Bart shook his head slightly, chuckling in wonder. "Then it's true that the bodies we are building are better than human ones."

Paul moved his chair a bit closer to Bart's. "Actually, Bart, I was wondering about just that. Do you think we should let people who transfer their personality data improve their 'copy'?" Paul asked, making air quotes. "After all, 'Improve' is Eidyia's motto!"

Bart's eyebrows furrowed. "What do you mean?"

"Well, say, for example, I was making a copy of myself. I might want firmer muscles or longer legs."

"I guess that will be for the Committee to decide." Bart was referring to the four senior executives, including Paul and himself, who supervised the transhuman project. "As you know, I prefer to be selective as to who is allowed to buy our service, just like we discussed, and not let people change who they are when they create their clone."

"We still must come up with rules as to who should be allowed to buy a transfer. We agreed that we wouldn't let just the ultra-rich buy them, but we never really discussed how we'd get there."

"True." Bart sighed. "Honestly, I think about that question pretty much day and night. When did we plan to have that discussion? At the next retreat, correct?"

"Yes." Paul put his cup down, got up, and walked to the window. He would need support to get things done the way he wanted, and he knew someone who would be perfect in that role. "Do you think we could invite Christine?" he asked carefully.

"To our retreat?"

"Well, yes. It is a big issue. We will need all hands, or all minds, on deck."

Bart gave him a skeptical look. "Well, all hands, maybe, but she's not on the team, Paul."

"She could be. She's the best in her field, as you know. And a good friend. I mean, of both of us."

"Well, maybe more a friend to you than me," Bart said with a wink. He walked over to the coffee machine and pushed a few buttons. The machine's grinder began to whir. "I guess you're right that it couldn't hurt to have a legal expert weigh in on this."

Paul smiled. "I'll see if she's available."

***

Paul opted to call, which was a rarity because almost no one made phone calls anymore. Christine didn't pick up, so he left a message and waited to hear back. When the phone rang an hour later, Paul answered almost immediately.

Before he could get a word out, she said, "Hi, it's Christine." This was not exactly necessary because Paul's watch had already said "incoming call from Christine," but old habits die hard.

"Hi, Christine. Thanks for calling back so quickly."

"I must say, I'm surprised, Paul. A call from you is unexpected, to say the least." Her voice sounded guarded, almost suspicious. He couldn't blame her.

"Well, here's the thing. I'm working on a big project at Eidyia, and we're putting together a small team to go over some sensitive parts of the project that could benefit from your professional expertise. To make a long story short, we're organizing a meeting to discuss the project, and I, well, we, would like to know if you might be available."

"In Portland?"

"No, in Jackson. You know, my house there."

"Ah, I see. So, you're inviting me to this meeting?"

"Not exactly just yet. But before I put your name on the list of potential invitees, I wanted to check if you would be interested, and available. All expenses paid, of course."

She was silent for a moment before finally saying, "I'm down."

Paul smiled. "Excellent. I was really hoping you would say that. Let me talk to Bart and the gang and I'll be back in touch in a couple of days."

CHAPTER 5

On the treadmill at the gym on Thursday afternoon, Christine started listening to a new biography of the Russian poet Sergei Yesenin, who had produced so much in so little time—his life interrupted by suicide at age thirty. The book was full of pictures, and she could see on the treadmill screen how so many of his verses matched the views from his homestead. The valley, the river, and, of course the birch trees that had inspired one of his most famous works.

Christine was genned up on all things Russian. The real Russia that had produced so many masterpieces before the Bolsheviks and later the Soviets destroyed its artistic potential by imposing official art, an oxymoron if there ever was one. Her mother had been a Russian literature professor at Yale, and Christine had inherited her love of Russian poetry and cinema. She had read almost all of Pushkin and many others, including Yesenin. Of all the poets in human history, Russians had the fiercest pen—before the Bolshevik takeover; not those lackeys who wrote mildewy verses to please Stalin. There were very few nooks and crannies of the human soul that true Russian poets had not explored in depth, and the poems that emerged from their inkwells created new understandings. Reading them was like discovering a new world, like when you learn a word in another language to describe a reality that would necessitate a whole sentence in your mother tongue.

In her twenties, Christine had accompanied her mother on two long trips to Russia, where she had insisted on making the journey to Ryazan to see Yesenin's homestead. She knew many of Yesenin's poems by heart, including "The Birch," her favorite for its unnerving purity:

And the birch tree stands
In sleepy silence,
And the snowflakes are burning
In a golden fire.

Oh, how she wished she could read it in the original language! She had tried to learn Russian and managed to reach what she called "proficient tourist" level, but she was nowhere near fluent enough to parse the sensual connections between the phonemes and the emotional punch they packed. How grateful she was for human translators. True, AI was reasonably good at translating works in prose, but poetry is not a prose by any other name. Poetry proved that AI has limits, because non-human translations were incapable of conveying the multiple layers of meaning embedded in good poetry.

Back at her townhouse and freshly showered, Christine asked Maya to play Tarkovsky's *Stalker*.

Within a few seconds, the MosFilm logo appeared on the screen. She had seen the movie at least five times, and each time she rediscovered it, she saw something new. Was this not what a voyage inside's one soul might look like? Men in a prohibited "zone," looking for a room in a building in which their deepest wishes can become reality, but to get there, they must go through dark, dangerous, and creepy places, like an old sewage tunnel. And when they finally get to the room, they're no longer sure they know what their deepest wish is. Are we not like that? Chasing after something we think we can identify, only to realize, when pushed to put our finger on it, that we have been chasing an evanescent target, a cloud of smoke, an illusion? The realization can be liberating, bringing us back to the here and now. But it can also be a powerful pull toward deep depression.

When the movie was over, Christine just sat there for a few minutes, still feeling it on her skin, in her bones, and echoing everywhere in her mind. The voiceover verses from the Tyutchev poem at the end of the film, read by a little girl, had managed to unsettle her once again, just as they did every time she heard the Björk version:

> *I love your eyes, my dear*
> *Their splendid sparkling fire*
> *When suddenly you raise them so*
> *To cast a swift embracing glance*
> *Like lightning flashing in the sky*

Yes, the power of a lover's glance, and desire, and love. All things that were not exactly abundant in her life these days.

She toyed with the idea of playing another movie, perhaps *Andrei Rublev*, but decided against it despite the attractiveness of the masterful trip through medieval Russia, with its magnificent icons as illuminating milestones. Two Tarkovskys in one day was simply too much for anyone's soul.

Instead, she poured herself another glass of grigio from a little town near Venice she'd visited with Paul and asked Maya to take her to the Eikasa home screen.

The logo of Eidyia's video streaming service popped up and a voice said, "Hello, Christine. What would you like to watch now? I have new recommendations for you."

"Something about love," she answered.

From her S-Chip, Eikasa knew that Christine was one of the rare viewers who didn't like happy endings. Years of watching Russian films and reading Russian literature will do that to a person.

Knowing her preferences, the voice said, "How about *Empty Quarter*?"

"Hmm, is that an Italian movie?"

"No, French. 1985," said the voice. "The director is Raymond Depardon. I think you will like it, Christine."

"Okay then, let's try it."

Eikasa's algorithm had done it again. The long, slow, cinematic trip from the coast of Djibouti down the Nile was just what the doctor ordered. Was that long, lonely journey through Africa not like her own love life?

If she was honest with herself, she still had strong feelings for Paul. She couldn't shake them, and worse, they acted as a shield against other possible relationships. Christine and Paul had shared the same freshman dorm at Vanderbilt—East House, whose four huge columns desperately tried to make the visitor forget about the decrepitude of the 1920s facade. Christine had been sitting on a rocking chair on the porch reading on her tablet to prepare for class when she'd first noticed Paul approaching on his motorized skateboard wearing huge, black, fuck-you headphones. When he wasn't listening to music or a podcast, he'd wear them around the top of his head like Caesar's laurel wreath. He managed to straddle adroitly that border between freedom and antisocial behavior in a way that filled her with envy—she could at best occasionally manage a light showing of irreverence. She had a deep love-hate relationship with her own strong sense of discipline, always picking up when her behavior might ruffle feathers, getting her kudos for being such a "nice person," but at the price of so many regrets. Perhaps she'd believed that being with Paul would inject a much-needed measure of nonconformity into her veins.

They had stayed together for a few years, but progressively drifted apart as Paul invested all his energies into Eidyia. The breakup had been half baked. They had just crossed that tipping point past which staying together no longer weighed as much as being apart, when being alone suddenly makes more sense than trying to be one with someone else. But it was one of those breakups that leave lingering regrets. Christine had loved Paul, or at least she thought she had, at some point—and maybe she still did. Love is a mysterious substance that the rational brain cannot penetrate, a flow of energy greater than will itself sometimes, yet impervious to reason. Christine should have stayed away from a man who could be sweet when he wanted but used people as stepping stones to cross whatever crosscurrent life put in his path, even if that meant keeping other people's heads under deep emotional waters.

But apart from a rather disastrous one-night stand four months ago, her lingering feelings for Paul had kept her away from romantic encounters, and she'd basically been alone since they split up the year before her mom's death. Thinking about her life that way made her sad, but as she was just about to finish her third glass of grigio, the movie ended in Venice, making her smile.

"Next time I watch a movie, I'll *stay* in Venice," she told Dewey, who was curled up next to her on the couch. "Visconti is what I need." In her mind, scenes from *Death in Venice* played, and the slow, long waves of Mahler's music, the adagietto from his Fifth, rippled through her body, giving her goosebumps.

Her watch buzzed, and three purple flowers appeared on the screen. She and Rachel had a code: they sent flower emojis when they wanted to speak, and the other would call back whenever it was convenient. A nice way to avoid having to hang up, which left a weird feeling, especially when the caller is your best friend.

Christine smiled. "Maya, call Rachel."

Moments later, her best friend's pale face appeared on the screen, grinning like the Cheshire cat. "Hi, Christine. I was just watching a movie and it made me think of you."

"Well, that's funny, because I was just watching a movie that made me think of you! It was filmed in Africa, your favorite spot on the planet."

"Synchronicity, they call it. I love how this keeps happening, like we're connected somehow. It's so cool. So, how are you?"

Christine took a long path through the labyrinth of random events in her life, but in the end, she told Rachel about her empty love life, and then Paul's call.

"Paul?" Rachel's voice dripped with disapproval. "Hmm, you may want to be careful here." She'd always thought Paul was just using Christine, and other people.

"I know, but…"

Rachel eyed her skeptically. "That *but* could be a source of a whole lot of trouble."

Christine knew deep down that she was right. But love is a cunning little thing. It puts our thinking in a safe deposit box deep in the mind and throws away the key.

***

The next morning, Christine decided to walk to the law school. The sun was shining, and the weather just right. Happiness bubbled up inside her. She had overtaught for a few years, so the dean had reduced her teaching load, and having just one course to teach this semester meant time to write and daydream and pursue other passions.

As she approached the building, she noticed a guy walking towards her, and when they passed each other, she thought for a second that he was looking disapprovingly at her scarf, a Gucci square that her mother had brought back from a trip to Italy. As soon as she got to her office, she took the scarf off. Maybe it didn't fit after all.

She started class with a video she thought would be a good follow up to the previous class's discussion. It was Safety Network footage of an actual arrest.

As the video began, the deep, loud, masculine voice of a robot police officer said, "Put your arms up and drop the weapon!"

Visibly sweating, the would-be thief turned and shot the robot. The 9mm round barely scratched its armor, and the robot retaliated with a powerful, Taser-like jolt that knocked the thief on the ground. Within twenty seconds, the robot had tied the person's hands behind their back, then it carefully picked up the weapon and took out the magazine. A minute later, the would-be thief and their gun were in an armored police version of the PC, on their way to the station. The thief's S-Chip had sent a signal that the individual was trying to enter an apartment where they had no known connections, through a window, and alerted the police.

"Another victory for the S-Chip," the voiceover added, as the thief was shown being driven to a special area behind the police station.

Christine turned off the video and turned to the class, firing her opening salvo. "Do we have any reason to worry about discrimination with AI? How can we prevent discrimination in the use of AI systems and police robots, and in AI decisions on bail and sentencing in particular?"

No one answered.

"Mary-san, let me start with you today."

Mary hesitated. She was normally the biggest fan of progress in robotics, but Christine had assigned several articles about bias in the criminal justice system that she could see had got Mary thinking. "Perhaps," she said finally, "there's a way to program values into the system?"

"Values?" Christine's eyes opened wide, and she tilted her head, intrigued. *Not what I had in mind, but let's see where this goes.* "What kind of values?"

"Fairness, for example," Mary said in a low voice.

Christine took a step forward. "How would you define it?"

"Well, people know what's fair even if they don't always act or play fair."

"Are you so sure there is a universal definition of fairness?" Christine asked before turning her attention to the back row. "Jerry-san, what do you think?"

Jerry jumped, and his face started to redden. "Hmmm, fairness, well, I'm all for it."

So much for him getting more comfortable speaking in public. "I guess we all are, but do you think we should program it into robots?"

Jerry shrugged. "Sure. I can't see any reason not to."

"I'll give you one," Mira jumped in, fixing Jerry with a challenging glare. "Because we can't. Robots are just data crunching *things*. And fairness is not data."

"Well," Christine said, "that may be right. But if aliens landed here tomorrow, do you think you could explain fairness to them?"

"How do you explain Rawls to an alien? Or to a robot?" Mira retorted.

"Rawls? Yes, there's one good example." All the students had read—or were supposed to have read—a book by the famous American legal philosopher John Rawls in their previous year. In *Justice as Fairness*, Rawls argued that cooperation and equal rights for all would provide the kind of structure that makes a society fair and just.

"Can we program Rawls into a computer? Squeeze fairness in code?" Christine asked.

A long silence ensued, but that was to be expected. Silence meant that at least some of the students were doing some deeper thinking about the question, and that had value.

"I think," Jorge said, breaking the almost meditative mood in the room, "that fairness, or maybe better unfairness, can probably be translated into data. What if you took thousands of situations and asked people to rate them as fair or unfair?"

"Interesting," Christine encouraged. She liked Jorge and knew how hard he'd had to work to get into law school after crossing the border with his family illegally at age four. Somehow that officially made him a "Dreamer," though his life had been anything but. Jorge's grandparents had been deported and died in Honduras before he could ever meet them. He'd told Christine that he planned to use the power of the law to reduce racial inequalities, especially those caused by AI. He might have a thing or two to say about fairness. Christine moved closer to him. "But how do *we* decide what's fair, Jorge-san?"

Jorge was thinking about his answer when Weijia jumped in.

"Didn't Rawls himself define fairness?" she asked. "Something about equal opportunity and providing a boost to the least advanced members of society?"

"Yes," Christine said, disappointed that Jorge hadn't finished his thought. She made a mental note to get back to him. "Fair equality of opportunity. And let's not forget equal rights and basic liberties. I'm not sure that's a definition you can feed into a machine, but it's a good start. Do you think we can operationalize it, Jorge-san? These can be good guideposts for humans, but can we use those same guideposts for AI and robots?"

Jorge frowned, but there was a question on his face. "Couldn't we just find data, like real world data, on how humans use Rawls's principles?"

Now, there's a true computer science answer.

"Yes, sure, like, these robots are only able to do one thing, and that is crunch data," Mira jibed. She moved her pink hair behind her right ear with her hand as she always did before trying to score a big point. "Do we believe even for a minute that fairness is about *data*? For one thing, we don't have the right 'dataset,'" she said, making air quotes, "and we still make those calls one by one. Psychologists have shown that young children know what is fair or not on the playground, and they're not crunching data. This is not about data."

Jorge raised his hand, and Christine smiled. Mira wouldn't just walk away with that one. "Yes?"

"What I meant," Jorge said, turning to Mira, "is that the data is the data. Those kids you're talking about, they *do* process data. The data is what they experience every day on the playground or at school. It all depends what data you use and what you do with that data. There *are* definitions of fairness, and that means they can be programmed into AI systems."

"Can you remind us, what those definitions are?" Christine said, to help him make it 2–1. "I'm not sure everyone is familiar with them."

"Sure. So, one is to ensure that machines treat every group in the same way. I mean, you can compare outcomes by group, whether it's by gender, race, or any other set of criteria. If the group is large enough, outcomes should be similar. If members of a particular group get longer sentences, or less chances to get a job interview when resumes are processed by machines, then there is some evidence of unfairness."

"Fewer chances," Christine said with a smile. She couldn't resist correcting students' grammar but felt a pang of regret when Jorge's smile faded. She had taken his victory from him. Maybe she could still make amends. "Thanks for that, Jorge-san. Actually, as we saw in the readings for this week, the risk is that when machines learn from historical data, they just perpetuate historical biases. It's not bad intentions on the part of programmer or machine, but pure data crunching reinforces existing biases. That can be used for good. The data can be analyzed, and in fact, the whole process can bring those old biases to light."

You're in a hole. Stop digging, she told herself as she walked across the room. "Okay. Let's shift gears a bit and discuss codes of ethics. Are they a good way to achieve fairness? Tommy-san, what do you think?"

"That's a tough question." Tommy said. "I guess I would ask first, what can those codes do? Maybe they are just window-dressing, like many people say." As he spoke, Esther's eyes were firmly planted on his back.

Are they? Christine thought, wondering idly if something was going on between the two of them. She brought her mind back to the task at hand. "Have those codes worked up to now?"

"Some have, I guess. The codes say human tissue cannot be used to build robots, and the only known attempts to do this were met with huge backlash."

"You mean like the company in China that was using actual human skin?" Christine asked.

"The company was called Jiancheng," Weijia supplied.

Tommy nodded and continued, "Then the code of ethics on sexbots is also pretty much applied, like the part that says sexbots cannot be made to look like minors."

"Well, now, that's also, like, the law," Nadia cut in.

"True," Christine added, "but I think Tommy has a point. That law was basically a copy of an industry code."

Tommy turned towards Esther and smiled unmistakably. *Shit, I missed that one all right.*

"As an undergrad in robotics," Roger cut in, "I remember reading an article about how robots and AI in general must not just be intelligent but beneficial. Some Berkeley professor, I think. His point was that 'intelligent' means that AI should find ways to meet its objectives, but 'beneficial' means that AI must be able to fulfill *human* objectives. Or something like that, anyway."

"But that would mean treating AI systems as slaves," Charles protested.

"Doesn't that assume that robots should be treated as humans or some equivalent of human?" Christine asked.

"What do you mean?" Mira asked, looking genuinely perplexed.

"I disagree," Mary said, predictably. She leaned forward. "I think the problem is that it assumes robots somehow understand what an objective actually is. If we program a robot or some other AI system to achieve a task, it just follows a path. It's trying to get from A to B, but it does not see B as a destination, and when it reaches B, it stops or moves on to another task. It must have some sort of conscience to realize that it has objectives, and it does not. If it did, it might wonder why it must follow our objectives rather than its own."

"Doesn't that just assume that its objective is different from ours?" Roger asked.

"That's exactly what I mean," Mary said, perking up. "When people are asked what objectives they have in life, they may say things like 'I want to be happy,' or 'fulfilled,' but the reality is that what they actually *do* doesn't fit that objective. People might *say* they value cooperation, but in reality, people prefer to compete. People might *say* they want the government to protect the environment or reduce inequality, but it took two back-to-back category five hurricanes that put most of Miami and the South Carolina coast under three feet of water before anything happened in Washington, and there are still climate change 'skeptics.'" Her mouth twisted into a disgusted pout, the kind you get when you know there are billions of people who won't be swayed by any logic or fact. Her cheeks had turned rosy with the heat of her passion. "We always say one thing and do another. AI systems are not like us. They learn by processing facts, not some made up crap. With the S-Chip, Eidyia knows what we really think, and what we *really* want. Do you see what I'm getting at?" Her eyes were the size of golf balls.

"I think so, Mary-san," Christine said, trying to bring the class back to earth. "Psychologists have identified those biases, including some inability to accept facts that contradict a belief, so yes, I think I get it."

Mary sat back and dropped her shoulders.

"It can get much worse, actually," Roger said, earning him a smile from Mary. "Try to explain to an AI system that, like, some people don't believe in, say, climate change, now that we're heading for a minimum 3.5-degree increase, or that the earth is flat—because apparently there are still some dickheads out there who think this is a *fact*—then the AI has two choices, and they're both bad. Really bad. If it's fully subservient to its owner, then it must integrate this non-fact into its programming and act accordingly. So, choice one is bad. The second choice the AI has is to realize the owner is mentally deficient. What does it do then? Send you straight to the hospital for a psych hold?" Roger concluded, opening his hands to emphasize the point.

"Wow, I never saw it quite like that." Jorge tilted his head from side to side. "But it does seem to make some sense."

"It's easy," Roger continued. "AI works with reality and facts, and we don't."

Mira jumped in, pushing her pink hair behind her ear. "But AI can also be used to manipulate people. Machines do that all the time. Getting people to buy stuff they don't need or vote against their own interests."

"That," Roger said, "is actually my ultimate nightmare."

Christine suppressed the urge to roll her eyes. *He uses 'actually' way too much, the little know-it-all.*

Roger continued. "If the machines realize we are just a bunch of easily manipulatable morons, what is our future? Take climate change again. An AI system in charge of, say, national defense could see climate change as a major threat and want to take measures, and make a list from, say, A to, I dunno, K. We might tell the system that, well, you know, we can't cause this or that bankruptcy so we cannot take steps A, C, D, E, G, and K on your list because the profits of some oil company will go down and Wall Street won't be happy. An AI would easily see that as irrational. To put the paper value of a company ahead of the protection of the, like, the only planet we've got, you know?" Roger paused and looked around the room. All eyes were on him. "It's like sleepwalking towards a precipice. Then if someone were to say, well, there is no such thing as human-made climate change, the AI system would think that person is absolutely nuts. It's like saying water is dry or the sun won't set tonight."

Now Christine found herself surprisingly agreeing with him. Yesenin popped into her mind:

And the charlatan, the murderer, and the villain
Whistle like autumn across the entire country...

She forced her mind back to the discussion at hand. "I see. So, the idea is that AI machines are better than humans?"

"What I'm saying," Roger clarified, "is that they don't think like we do. They analyze *data* and take a probabilistic approach to decision-making. A human decision is a complex mix of stuff. It includes maybe some rational thinking and data analysis, but also hormones, emotions, cognitive biases, neuroses, and so many other factors. Think of how the body reacts to thoughts. Your feel vertigo when you imagine yourself staring down from the roof of a tall building. Your heart rate goes up. As if thoughts were real."

Now something in Christine's brain summoned a line from Turgenev: *We sit in the mud... and reach for the stars.*

"I can see how machines don't think like us, Roger-san," Christine said. "I'm not sure why we would want them to. This idea of creating neural networks to copy the human brain hasn't worked. Why wouldn't there be multiple ways of 'thinking'?"

"Yes, you're right, Professor," Mira agreed. "But machines can use data about us, about everything we do, which almost everyone gives them with the S-Chip, and then they can manipulate us into thinking this or that way. Wasn't it shown that AI bots got more than one fifth of US Congress members elected in 2032?

They didn't pick a candidate for this or that reason. They were instructed by humans to manipulate voters, but what if they did pick on their own? What if they had…" She lowered her voice. "An agenda?"

"I'm not sure I follow," Christine said. "An agenda to do what?"

Mira looked down at her tablet. *She's given up on that one.* No one spoke. Christine felt a big drop in the energy of the room and didn't need to turn around to know that the fifty minutes were almost up. Then Nadia raised her hand.

"Go ahead, Nadia-san."

"I read this article about using machines to run the energy grid according to evidence-based goals. If they did, they wouldn't like let humans decide much because collectively we're not good at anything: environment, inequality, you name it. The good thing is, right now AI doesn't, like, *want* anything. It just *does*." Nadia looked down at her laptop, as if to say, 'now leave me alone.'

Christine picked up on it and moved on. "Much to think about, Nadia-san. Please send me a link to that article and I'll circulate it to everyone."

She looked around the room. Many students looked shellshocked. Some classes do that, but in her experience, it always sunk in eventually. *Time to let time do its thing.* "See y'all next week. Come to my office if you have questions about the stuff we talked about today."

As she walked back to her office, Christine's watch buzzed. It was set to block all notifications except from Paul, so she knew it was about the Jackson Hole meeting, which he'd confirmed that morning.

"Maya, what's the message?"

"Your flights to Jackson have been booked. You leave next Saturday at 11:35 a.m." Details about the flights scrolled on the small screen.

Christine wasn't sure what she was looking forward to the most: seeing Paul again or hearing about his new project. That simple message had opened a tap of uncontrollable thoughts and emotions about her mom, Paul, love… As memories of being with Paul in Italy flashed through her mind's eye, she entered her office and sat, looking pensively out the window at the falling rain. She'd never liked rain, but Paul had once told her there was much to like in something that washes the world clean and helps so many things to grow.

Then a student knocked on her door, bringing her back to the mundaneness of the here and now.

CHAPTER 6

As they exited the classroom, Mary asked Jorge if he wanted to grab a coffee. They made their way to the small law school café, ordered their oat-milk cappuccinos, and went to sit outside on the adjoining terrace. A big oak tree provided shade, its ageless trunk giving a sense that one was rooted.

"So, what's on your mind?" Jorge asked.

Mary immediately launched into a monologue about Mira that culminated in, "Why be afraid of robots? They just make life easier." She glanced at Jorge. "Sorry, am I boring you? I guess I have a lot on my mind."

"Not at all. I was just thinking of my family in Honduras. They're coffee farmers in a region called Copan, high in the mountains. They send me a big bag of coffee beans now and then. It's so good."

"How did your mind go from my rant about Mira to the mountains of Honduras?" Mary asked with a puzzled smile.

Jorge met her gaze. "My family used to do a lot of the work manually, but now all of it is done by robots. When I speak to them, they seem bored. Money is good, but they have nothing to do. They watch Eikasa all day. So, I know that robots make things easier, but I'm not sure easier is always better."

Charles and Nadia exited the café and approached their table.

"Hey folks, can we join?" Nadia asked.

"Of course!" Jorge said with a genuine smile on his face, motioning towards two empty chairs across the table.

Charles set down the tray he was carrying, which held two bagels covered with copious amounts of cream cheese, one of them with a slice of pale smoked salmon on top.

After an awkward ten-second silence, Nadia asked if they were interrupting, and Jorge gave them a short summary of the chat they'd been having.

"I guess it's like a muscle," Charles said, picking up his bagel.

"Not sure I follow." Mary took a sip of her cappuccino.

"You know, the less you use it, the smaller it gets," Charles said when he finished chewing. "The risk is that robots will do everything for us. And we'll lose not just muscles like Jorge's relatives on the coffee farm, but our brain muscles too."

"Don't tell me you guys agree with Mira and all this anti-robot shit," Mary said, tilting her head back.

Nadia looked at her, pursing her lips. "I don't think it's, like, anti-robot to be pro-human. After all, we keep building more and more robots to do more and more of the things we used to do as humans without really thinking where this train is going, that's all."

"Look," Mary said, her expression now mildly aggressive, "robots do a lot of things better than we can. Look at the number of new drugs machines have discovered. Why would we want to prevent that?"

"I don't think that's what I said, Mary-san." Charles smiled disarmingly. "I was just saying that we can also think of the future for humans when machines do everything. They write contracts, predict the outcome of cases, and so much more that even lawyers will be out of a job soon."

"So?"

"Well, you're in law school!" Charles still had that winning smile on his face, as if he had just scored a touchdown.

"Yeah, to learn robot law so I can work *with* robots, not against them!"

"Folks," Jorge jumped in. "Take a deep breath." He raised both hands, palms down in the universal sign of appeasement.

After a few seconds, Mary met his eye. "Sorry, I guess Mira just got under my skin more than usual today."

She felt doubly bad because she really liked Jorge, and now she had kind of made of fool of herself. Besides, he had backed her up today. Why had she lost it like that?

Charles took another bite of his bagel, his mind still savoring the win. *A law student arguing for robot lawyers? Work with robots, not against them? How plain fucking dumb.*

CHAPTER 7

The flight to Jackson was surprisingly quick. Christine barely had time to read a few law review articles and sip a bit of third-rate chardonnay. When she exited the airport, a PC was waiting with her name flashing on the windshield. She was still impressed by the technology that allowed a windshield to act as a screen to display messages like this even though it had been in use for many years now. Part of her list of small things she always found surprising, like how you could send a letter, now that those were back in fashion, from a mailbox on a rural road in Maine to a friend's condo in Los Angeles for just $1.95. All the steps that small envelope had to go through. It was amazing if you really thought about it.

After a quick twenty-minute drive from the small airport with its walls covered in dusty stuffed moose and bison heads, she arrived at Paul's mansion. The heavy iron gates opened automatically, obviously programmed to recognize her S-Chip, and the PC drove in. Christine's heart pounded in her chest as she exited the car and approached the front door.

The door swung open before she even had a chance to knock, and Paul stood there, smiling disarmingly. "Welcome. Did you have a good trip?"

She could feel her insides melting and desperately tried to calm down, cursing that part of her that still found him irresistible. "Uneventful and very quick, I must say. I guess that means good. Thanks for arranging a direct flight." She hoped he didn't notice the vibration in her voice or see through her business-as-usual act.

"Of course! I—we—are so happy you could make it. Come in!"

Hmm, no hug? Disappointment crashed over her. *This really must be all about this secret project.*

As she entered the mansion, she was reminded of Paul's eclectic tastes. The modern furniture clashed with the huge Murano chandelier that Paul had bought when they were on the eponymous island just off Venice. She still remembered how they'd hugged on the vaporetto that took them back to the Carlton Hotel on the Grand Canal. The world had belonged to them then, her love for him as strong as ever.

31

Paul set her travel bag in the corner and gestured for her to follow him to the massive kitchen. It was an impressive space, featuring a 60-inch range with six gas burners, two electric cook plates and a grill, a small pizza oven hanging from the wall, and all topped by a huge metal hood that looked like it could fight a hurricane toe-to-toe. The azulejos Paul had specially ordered from Portugal gave the kitchen a je-ne-sais-quoi air of homey professionalism.

"Would you like a cup of coffee?" Paul asked, gesturing to a gleaming Simonelli that hadn't been there the last time she was here.

The question put an abrupt end to her daydream. "Yeah, sure. One cream, as you know. Thanks."

Paul pushed a few buttons and the machine started grinding. Moments later, he handed her a big mug with a picture of the Rialto in Venice, and she remember how she'd convinced him to spend the 35 euros at one of the many tourist-traps in tightly packed rows on the bridge. The colors painted on the mug had rapidly faded in the dishwasher--a symbol of her fading love for Paul? The thought reignited her mental trip back to the City of Doges, but Paul interrupted her again.

"The others are already here. Shall we?"

"No coffee for you?" She raised an eyebrow. Paul was such a coffee addict.

"Later. I just had one with Bart."

"Oh, it's been a while. Where is he?"

"With the others, in the vault downstairs."

Vault? She had never heard of any vault. Christine followed Paul down a steep flight of stairs from a small storage room behind the kitchen into a long, windowless corridor. They stopped in front of a metal door. Paul put his watch against a panel on the wall, which opened to reveal a number pad on which he typed a nine-digit code.

"Welcome, Paul," said a soft masculine voice.

"Two of us," said Paul. The door opened, and he gestured her forward. "After you."

Christine stepped into the room and took in her surroundings. The walls full of screens and discreet lights ensconced in the ceiling made it look like a high-tech boudoir. In a space carved in the grey wall stood another shiny Italian coffee machine. There were five white leather chairs with the latest model Apple tablet in front of each one. Three of the chairs were already occupied, and she immediately recognized Bart. He hadn't changed much. Still kept his head shaved.

"Hi Bart," she said, smiling broadly. Bart didn't say anything but looked at her and pretended to doff his nonexistent hat. "Good to see you. You look great!" After an awkward pause, she remembered that Bart didn't respond well to compliments. Or was he just already focused on the business at hand? Yes, typical Bart. "Why all the security?"

Bart smiled. "You will understand the need for security in a few minutes. By the way, your S-Chip and watch will not be able to transmit from this room. In case you are expecting messages or calls."

"Oh, this sounds serious!" she said brightly, trying to lighten the somber mood.

"Well," Paul said, "let me introduce you. You already know Bart of course. This is Koharu Tanaka. They're our lead biologist on the project."

Koharu had even shorter hair than Christine, wore no makeup, and was dressed in the androgynous wardrobe that had defined Silicon Valley since the sex and pay inequality scandals of the 2010s. They wore a white shirt with a thin black tie and dark brown Blundstones. Quite the contrast. Like so many people, they had chosen to move past the arbitrary, culturally defined gender binary, embracing mixed fashion to bolster pluralism.

Paul continued. "Next to Koharu is Jeremy Blakes, our chief engineer. He's working on stuff you'll probably find amazing, but I'll let him explain."

Jeremy looked like someone who had permanently replaced sleep with increasingly incredible amounts of caffeine. He hadn't stopped fidgeting since she'd entered the room. He smiled at Christine in the way that children do when parents ask them to be polite.

Christine smiled at them. "Nice to meet y'all." Growing up in Huntsville, Alabama, had left the word stuck to Christine's tongue for life, and she'd found that the chumminess usually softened coastal elites a bit, but not today.

"So, Christine, I'm sorry to start on this note, but I must ask you to sign this NDA." Paul put a short document in front of her and laid a pen on top of it. "I hope you don't mind."

Old style, pen and paper. Christine raised an eyebrow, but quickly scanned the document, signed it, and returned it to Paul. "Sooooo, what is this big project?"

Bart caught her eye. "Not what you expect, Christine, whatever it is that you might expect. We are about to finish a prototype of a new kind of robot."

"Something like the R3?" she asked. Almost everyone was familiar with the now-common Eidyia R1 and "special purpose" R2.

"We call it the R-H, actually," Koharu said.

"H?" Christine frowned in confusion.

"Yes," they confirmed. "H as in Human."

What did *that* mean? "Ah, you mean this new robot will look more like a human?"

"More than that," Bart said. "It will essentially *be* like a human."

Christine felt like she was swimming against the tide towards a distant shore of understanding while everyone else was surfing. An uncomfortable posture. "What do you mean *be* like a human?" she asked, picking up her Venetian coffee mug.

"Well, it will look like a person," Bart explained. "It will be able to see, hear, and feel like a person. In fact, better than humans. It will even be able to have sex."

The image of a lifelike sex doll popped into Christine's mind. "Wow! I can't believe it. Has AI progressed that much?" she said doubtfully. Then she thought, *Robot sex...is that progress?*

Jeremy cut in. "It's quite a bit more than that, Christine. We can actually take all we know about a person from years of data generated by the S-Chip and transfer a complete personality into a new body."

Although it felt like a train had just hit her, Christine tried to retain her composure. "But wait a minute, why would anyone want to do that and live with a copy of themselves?"

"That is one of the questions we plan to answer this weekend," Bart said. "We need to see what our market is. Or what we want it to be."

Christine took another long, slow sip of her coffee, only to realize it was getting cold.

Perhaps noticing her grimace, Bart said, "There are snacks and tea in the small kitchen there, and the coffee machine is here of course." He pointed. "Should we take a minute before getting started with our questions?"

"It's fine. I'm just a bit overwhelmed," Christine said. In her mind, she had just followed the rabbit down the hole into a strange dream world.

Paul looked at her, tilting his head and frowning a bit. "But you've been teaching robot law for years. I know you've thought about this kind of thing."

"Of course, Paul. I've read sci-fi books too! But having a theoretical discussion with a group of students isn't quite the same as seeing what people who actually build robots worry about. It makes it way more real. I feel like I just left the real world and entered a parallel universe."

"This project is *very* real," Koharu said.

Bart stood up, and everyone else did the same except Christine. They busied themselves, giving her a minute to compose herself. Paul left the room and brought her back a bottle of Perrier and a glass. Bart made another espresso, and Koharu refilled their cup of tea. After a few minutes, they all sat back down, and by some sort of silent agreement, all looked at Christine.

She took a deep breath. "So, tell me more. What exactly is it? Humanoid robots?"

"More like human robots," Paul said. "Or maybe I should say robot humans."

"What do you mean?" Christine asked, still uncomprehending.

"I mean that the plan is to transfer a person's, well, their personality into a robot that looks exactly like them," Paul explained patiently. "We are just about there. Skin, hair, voice, thoughts…"

Christine's head felt like a pressure cooker, questions pouring out of her mind faster than she could put them into words. Where to start? "Wait. Voice. How would you do that?"

"That's actually one of the easiest parts," Jeremy answered, eyes lighting up with excitement. "We have hundreds of hours of speech on file, generated by each S-Chip host, and all the interactions with their watches. Our voice algorithm can easily reproduce a person's exact voice."

Christine's mind shifted into analytical law professor mode. "And, Bart, did you say *thoughts?*"

"Yes." He fixed her with his direct gaze again. "That part was definitely trickier. We have to process tens of thousands of hours of speech, emotional responses, and so on from the S-Chip, and then we are able to create inferences, enough to build a total empirical model of a person. We can predict with more than ninety-eight percent accuracy what a person will do in any given circumstance. And of course, we implant all available memories as well, including all media related to the person."

"That redefines the expression 'photographic memory'!" Christine quipped as her mind finally began to slow. She uncrossed her legs and got up to make a fresh cup of espresso on a coffee machine she recognized. Familiar territory felt good right now. The machine used to sit on the kitchen counter upstairs.

"True," Bart said.

Jeremy looked like a school kid proud of his science project. "In fact, too precise a memory is not good if we want to replicate human personality. As you know, our memories vary over time, and that is part of the construction of our personality. We constantly rebuild our past to match our present."

"I know." Christine thought for a second about her paper on the difference between human and machine memory published in the *Vanderbilt Law Journal*, which lawyers now cited regularly to try to overturn jury verdicts. "I've read too many studies to count showing that eyewitness accounts are so often unreliable, very far from what actually happened."

"Exactly," Jeremy agreed. "That's why we recognize variance in the memories we implant, but I'm not sure it's much worse than actual human memories."

"Wow. Just wow." Christine said. "I can't quite get my head around all this."

She sat back down and held her cup up to her nose, trying to let it all sink in, as if the chocolate and berry aromas of the mountains of Honduras captured in the coffee beans could help her think better and see further.

CHAPTER 8

They took a break, which ended when Koharu returned from the kitchen with another cup of light green tea.

Bart looked around the table before his eyes settled on Christine. "So now you know that we have a technology that allows much of what makes a person a person to be transferred to a robot—one with skin, hair, etcetera that looks just like the person being 'transferred.' Now we need to figure out what the release of that information will do in the public sphere, and what market we may want to target. We brought you here because you are familiar with all the laws and codes of ethics for AI that have been proposed and adopted around the world. We need your input on the two questions I just raised, but also on possible changes to our current code of ethics because we have a game changer."

"Sure," Christine agreed. "First of all, let me say what I should have said already. I really appreciate the invitation. This is by far the most intriguing potential application of my research." She took a sip of her Perrier and looked around the table. She had their undivided attention. "As to codes of ethics, there have been so many. Where do I start? Initially, it was assumed that humans would program robots, so we could embed basic rules in their code. A bit like Asimov's obsolete Laws of Robotics. We now know that principles are hard to code into a machine, and then those principles often clash with the human behavior that AI systems observe.

"Machines program themselves now, as you know, so they can modify their code as they learn. The Second Montreal Protocol on AI of 2029, reflected in SAS1, is the only one I know that is applied almost universally. It contains a best-efforts clause for human programmers to make sure that robots will not cause harm to humans, and the obligation to install a kill switch, embedded in code that cannot be modified by the AI system. If, hypothetically, someone's personality was transferred into a new, inorganic—"

"Very organic," Koharu interrupted in a low voice. At that moment, they struck Christine as a mountain of quiet strength and determination hiding in a molehill of humility.

"Well then, a *synthetic* body, then none of this applies, it seems to me."
She felt as if she were in front of a classroom. Jeremy was typing on his tablet,
probably taking notes. "Let me just take two things that immediately come to
mind. First, you would have to decide if the transferee, if I can call it that,
has any kind of rights, or who knows, perhaps even human rights if it's a human
robot. Second, if this new entity makes decisions based on a person's previous
data but also new data it acquires itself, it is no longer the same person."

"Shit," said Jeremy. "Robots with human rights. Never thought about that.
But if they are human robots…"

"Exactly," Christine said. "As a matter of law, it's not all that clear what makes
a human human."

"Really? You would think the lawyers would have figured that one out by
now!" Jeremy's face expressed a dismay that Christine was all too familiar with. It
happened so often when explaining the intricacies of the law to non-lawyers. "And
isn't it obvious? It's our ability to think, correct? We can think, and animals can't,
right?"

She smiled. "I don't know whether you can say that animals think or not,
but no, thinking is not the deciding factor. I mean, as a matter of law. Say you take
someone who has an accident and is braindead or has suffered severe neocortex
damage. You would still call that person human, and recognize that this person
has human rights, though the law might entrust someone else to make decisions
on that person's behalf. So, thinking is not it. It's obviously not age, gender, or
race. It's not how we look either."

"Maybe it's the soul?" Koharu inquired pensively.

"Well, as a legal matter, I'm not sure how you would define the soul,"
Christine said, raising the pitch of her voice. "But it's true that many people
say something a bit similar. They point to our unique 'cognition.'" She paused,
as she always did when talking to non-experts after using a word that dug
deeper into the listener's brain, allowing time for their neurons to process
it. "I mean this peculiar mix of reason and emotion, genes and hormones,
nature and nurture. Language is another one that was used for years, but now
that machines speak almost like us and we've decoded fairly advanced linguistic
skills in certain animal species, I'm not sure we would want to use that as our
distinguishing feature."

"So, what is it that makes people human?" Koharu asked, letting go of their
calm harbor and looking increasingly perplexed.

"Biology, I guess. DNA." Christine shrugged.

"Hmm, but wouldn't that mean that an embryo is human?" Koharu asked,
moving forward on their chair.

"Yes," Christine said, "human, but not necessarily a human being."

Koharu frowned and their face tightened. "Human but not human being?" Christine nodded. "At conception, you have a bunch of human cells, but it's not a complete human being, at least not yet."

Bart also leaned forward, as if to occupy the table with Koharu. "I'm sorry, Christine. I don't get it. Our DNA is more than ninety-eight percent like some apes, right? Bonobos or something? So, basing the distinction on DNA cannot be right."

"It's tricky, I agree," Christine said, unfazed, "but maybe it *is* possible to draw a line. Maybe it has to be at 99.5, or 99.9 percent. I don't know. I don't really see another way to separate humans from other living things, though."

"Our new robot, the R-H, is made of synthetic organic matter, so it will not have human DNA," Bart said. "So, it won't be one hundred percent human. okay, but it will have a 'human' personality. How do we reconcile *that*—" He tapped the palm of his hand loudly on the table. "—with your DNA hypothesis?"

Christine took a few sips of water, considering her answer. "As I said, for the law this is unprecedented. Maybe humanness is a multifactor thing and personality is one of the factors. Let me flip the question this way: If I transfer person A's personality into a robot and then destroy the robot, is it murder? I think a court would say no. A is still alive. No human has died. It would merely be destruction of property. If you transfer A's personality into a robot, what happens to A? Do we have two As? Can the robot act as A's agent? You said it would make the same decisions over ninety-eight percent of the time."

"It's actually a bit more complicated than that," Paul said.

Christine, who had been looking at Bart the whole time, turned to face him.

"The way we see it, it's not clear that people would know they're not dealing with A, but instead with a robot into which A was transferred. Let's call it A Prime." Paul paused. "Like, who's to say I'm not a robot?"

"Very funny, coffee-drinking robot!" She paused, then said, "A Prime? Sounds like I'm back in high school math class."

"Well, if Koharu's personality is transferred into a robot, then there are, in a way, two Koharus—at least that's the way the rest of the world would see it," Jeremy said.

"Wow," Christine said. "I know I'm saying 'wow' a lot, but geez, y'all. How will the world work if everyone is there twice?"

"Well, the problem exists only as long as A exists at the same time as A Prime," Paul said. "A will get old and die someday. In most cases, A Prime will not, short of some freak accident. In fact, A Prime won't age at all. If they were to live side by side, it would be like living with Dorian Gray."

"But what you're doing here is kind of the reverse." Christine frowned. "The robot doesn't age, but it can see humans getting old around it."

"I guess that's right," Bart said.

Christine swiveled towards him.

"Sorry to insist, Christine," Bart said, "but do we *need* a new code of ethics?" She smiled faintly. *A tough discussion.* It was so unreal that they were talking about real robots. All her discussions in class had been so theoretical. Part of her mind was still trying to process the very existence of the project. She got herself back into gear and said, "If you're transferring 'humans' into robots, it would be like setting a code of ethics for humans. We have tried, arguably since Hammurabi, to get laws and rules of ethics in place. The link between the two is a great matter for debate."

"Hammurawhat?" Koharu asked. "I don't know this name."

"Sorry, I get lost in my own thoughts sometimes. Hammurabi was king of the Babylonians a long time ago. Eighteenth century BC I think," Christine explained. "And the Code of Hammurabi is a book of laws found inscribed in a stone stele. One of these is the famous 'eye for an eye,' the law of the talion. There are ethical principles embedded in the code." She smiled uncomfortably. "My point is, if a person is somehow copied into a robot, then the robot will act as ethically, or unethically, as the person whose personality was transferred. So, I'm not sure at that point you need a code, or at least not in the same way."

"Not in the same way? What do you mean?" Paul asked.

Christine turned her head. *He is not smiling at all.* She tried to let that thought pass. "Hmm, what you may need is a code about *who is allowed* to transfer their personality. I assume the R-Hs and the transfer process will not be cheap. Aren't you afraid of creating some sort of master race of the super-rich? After all, buying immortality has been on the lips and minds of Silicon Valley billionaires for at least thirty years. Tons of cash has been invested into solutions that were mostly meant to spend that cash, it seems to me. Until now."

"Well, there is a reason we are not in the Bay Area," Bart said. "I can assure you none of us are in this for the money."

Christine didn't miss the way Jeremy looked down into his mug.

"So, what are you in it for exactly?" she asked.

"I guess that's kind of what we're trying to find out," Paul said. "Technology always seems to come first. Then you see what happens when you deploy it. And a lot of it is unpredictable."

"I guess that's true," Christine said, "but you don't want to be told you built a new type of nuclear bomb, do you?"

Jeremy moved closer to the table, holding his coffee mug with both hands. "Actually, I think this can be good business, but it can also make the world a better place."

Christine, like everyone around the room, had her eyeballs glued to Jeremy as he continued.

"We're already making a few things better, like the skin, which will be much better at repairing itself than human skin, for example. So, we already accept that Transfers will be better in some ways at least. So why not do that on a larger scale?" He looked at Paul. "Remember how we discussed that when someone transfers, we can make them better in some way, maybe more ethical. Or even not allow bad humans to transfer at all."

Paul didn't reply, and the ensuing silence settled heavily over all of them.

As Christine was about to ask Jeremy to explain, Bart said, "I know you've had that thought in the back of your mind for a while, Jeremy. Let's get back to that topic later, all right?" He got up and stretched his arms in front of him, signaling that it was time for another break.

***

After another couple of hours of discussion during which Christine explained existing codes of conduct and ethics and what they usually dealt with, the group decided it was time for dinner. They went outside and sat around the fire pit, enjoying glasses of Vernaccia di San Gimignano that Paul had deliberately selected.

The wine, which had exceptionally floral aromas, would hopefully work its charm on Christine like Proust's madeleines. For Proust, the sweet childhood memories imbued in those cakes made them taste better, and Paul hoped this wine would bring to Christine's mind her first trip to Italy with him. They had done a tour of Tuscany, and on the road south of Florence had driven up a road and caught their first glimpse of the village of San Gimignano. Its twelve medieval towers anchored a town of low, ochre-colored stone buildings with red-tiled roofs, guarded by rows of cypress trees.

They had shared a wonderful dinner on the main square, the Piazza della Cisterna. Afterward, back at their hotel—in a renovated room in a fourteenth-century gothic building with a large window that filled the room with the rolling hills of Tuscany—they had made love slowly. It was possibly the best night either of them had ever experienced.

Paul's watch buzzed, and the computerized voice said, "Paul, someone is at the door with a delivery."

"Ah, the caterer must be here. Open the gate please, Sasha."

Paul's house robot soon appeared with the food on large platters, which it placed one by one on a big table near the fire pit.

"I hope you like vegetables and pasta," Paul said to the group. "The veggies are all local. Organic, of course. I'll join you in a minute. Just a few messages I need to send to people back at HQ."

By the time he returned, the others had finished eating, just as he'd hoped, and soon after, they began to disperse to their rooms.

Christine stood to follow the rest, but then turned back. "Can I help you clean up, Paul?"

"Oh, no, Sasha will take care of it." He pushed back from the table and gave her his most charming smile. "Would you like another glass of wine, Chrissie?"

He hadn't used that name in a long while, and it produced the intended effect. Christine's face melted like soft butter, and she moved closer to him. Then her gaze rested on the still-full plate in front of him. "Say, weren't you hungry? I don't think you ate anything."

Paul hesitated, then said, "You're right. I wasn't hungry. My mind is racing." He reached for the bottle of Brunello di Montalcino on the table and poured them each a glass, then steered her over to the chairs in front of the fire, where they sat close together.

Christine turned toward him, leaning in as she asked, "So, what's with the Russian name for your watch?"

"I thought you'd notice." That, too, had been a deliberate choice, another reminder of their shared history and interests.

"Well, yes, and the Tuscan wines too." She looked straight at him, as if building an invisible bridge that lips can then cross.

Paul wasn't in the same mood; he wanted to get to the point but didn't want to screw up. "I, hmm…sometimes I wonder why we split up. It wasn't all that bad, was it?" He took a sip of wine and moved back a bit.

Christine's eyes softened. "Not at all. Just the erosion of routine, like water on a rock. I guess the love boat crashed against the everyday."

"What do you mean?"

"It's a line from a poem by Mayakovsky."

"Oh, you and your Russian poets again!" he said with a smile, happy to be on terrain that was so familiar to them both.

"Russian poems can tell you more truth about the human heart in one line than entire books." Christine's voice was full of emotion, as it always was when she spoke on this subject.

"Is that what you see in Russian poetry? Why do you like it so much?" He knew the answer, of course, as they'd had similar conversations many times in the past.

"How would I explain it? It's like a tuning fork for the soul. The better the poem, the more powerful it is at aligning the various levels of the self, at making them vibrate at the same frequency." She became more animated as she spoke, her enthusiasm bubbling out of her. "A great poem can create one of those rare moments when you feel you've been touched by the grace of some timeless aesthetic energy, and all your cells have somehow been rearranged."

"Funny you said tuning fork, because for me, music did something similar, sometimes. Like some Bach cantatas."

"Yes, I can see that. Music is meant to play the keyboard of human emotions. Sometimes a single, simple note. Sometimes a more complex chord, creating more nuanced tones in the listener's palette. In rare cases, the music connects emotion and spirit." She looked at Paul. "The late Romantic composers do that for me more than Bach can. It's possible to achieve this kind of understanding with any form of art that looks like it was inspired by some force that transcends humans." She grew thoughtful, face lined with concentration as she pondered something.

Paul stayed silent, waiting for her to work through her thoughts.

Finally, she said, "What it is, I think, about poetry, is the ability of something that comes from all levels of the self to communicate with other humans. As a result, you understand things in a different way."

"That works for you, but I'm not sure it works like that for everyone."

"What about the poetry you wrote when we were in college? I thought it was very good. Remember how I urged you to get it published?"

"Yeah, and I did, but publishing poetry is a fool's errand. The market is so small. I think *Humanity's Eye of the Needle* sold two or three thousand copies. That's hardly a Faustian bargain worth selling your soul for."

"Selling your soul?"

"Yes. Publishing poems makes you, I don't know, vulnerable maybe."

"Well, I'm glad other poets took the plunge, money or no. Poetry has influenced the world a lot more than most people realize."

"Influenced? What do you mean?"

"Well, for example, Lucretius, a Roman from the first century BC, wrote a long poem called *On the Nature of Things*. It mostly disappeared for a thousand years or so, but when it was republished during the Middle Ages, it had a profound impact. It influenced Thomas Jefferson."

He listened with rapt attention, utterly mesmerized, as Paul had always been, by the brains and beauty shining in front of him.

"In a way," she went on, "I think that says something about us, as humans. Verses are not just milestones of human history. They're stepping stones. I don't think machines of any kind will ever get poetry. How could a machine understand how poetry is understood by humans? How many levels of the self can Baudelaire, Dickinson, or Yeast reach? How many of those levels does a robot have? If a machine ever gets poetry, then we're fucked. I mean, that should be our last chasse-gardée."

"Chass… what?" He knew what she meant, but he also knew that she liked to play professor.

"Chasse-gardée. A nice French word. It means a piece of land that is kind of a reserve. I mean the last human reserve."

"I'm not sure how much humans need a reserve from robots. Robots are just different, and I'm not sure they would want to be human." This conversation had gotten way off track. He leaned in. "Back to us, then. I guess it's true, Chrissie, that we let routine take over. And sixteen-hour days at work did not help." *It wasn't just routine, though.* "You know how you once told me, a month or so before we broke up, that life had taught you that luck means being *something to someone* in the maelstrom in which we must navigate every day? I still remember that, word for word."

"Being *anything to anyone* is the best most people get to, Paul. But yes, letting work take over was a huge mistake. The very idea of work-life balance makes no sense; no sense at all. What is work if life is the other pan of the scale? *Unlife?* But I wasn't a lot better than you on that front, I'm afraid, especially before getting tenure."

He looked at her in the eyes and grabbed her right hand softly, his fingers on hers. "Oh, I remember. You were obsessing over those articles and who would publish them."

"Maybe now that we're a bit older, we're also a bit wiser?"

"Hmm." Paul leaned over and put his lips on hers, gently, asking for permission. They kissed, softly at first, but soon all hesitation evaporated. "Let's go upstairs, Chrissie."

***

Back in Paul's room, with its reindeer chandelier and huge Italian four-poster bed, they looked like mishandled puppets as they tried to take their clothes off as quickly as possible.

"Let's do it in the firelight," Paul said. "I want to see your body with my hands."

"That's not like you, Paul. I'll miss seeing that furry chest of yours," Christine teased.

As Paul looked at Christine, what he saw in her eyes was a look of pure, undefended access to the soul. He pivoted slowly towards her back and put his face in her neck, where he started softly kissing her. Christine's head dropped. After long moments, she turned and pressed her lips to his neck. Too late, he realized his mistake.

"Oh Paul, you have a stain or something on your neck."

He jumped back abruptly.

"Paul, what the hell is the matter with you?" She looked genuinely puzzled, a 180-degree mood swing swiftly enveloping them.

"It's nothing. Sorry, Chrissie."

"Is it that thing on your neck? What is it, Paul?"

"Nothing. Well, nothing important."

"Come here. Show me." She went to the bedside table and turned on the light, beckoning him over.

"It's nothing, really, Chrissie." He could sense the mood was irreparably broken. *This is not going as planned.*

Christine continued staring at him expectantly.

No point delaying the inevitable. "It's a barcode."

"A what? A barcode, like on cereal boxes?"

"Well, something like that. I guess."

"Paul, why the hell would you have a barcode printed on your neck? Some new tech thingie?"

"Because I am not Paul, Christine."

Her jaw dropped. Like the floor had just been removed from all of her capacity to process information.

"I'm Paul Prime."

***

Christine screamed at the top of her lungs as her insides seized up. "Paul Prime! You fuck! And you were planning to have sex with me?" She grabbed the duvet and covered herself with it, collecting her clothes and pulling them back on awkwardly while trying to shield her body from view, her eyes silently screaming.

There was a knock on the locked bedroom door. "Everything okay in there?"

"All fine, Jeremy. Christine just saw a mouse," Paul Prime replied loudly.

Footsteps moved away from the door.

"A mouse! You fuck…" Christine's face was on fire, and her eyes felt like they were about to bulge out of her head.

Paul Prime tried to approach her, but she recoiled.

"I am sorry," he said. "I really should have told you, but then we thought this might be the ultimate test. So, you didn't know until just now, right?"

Christine glared at him. "Obviously not."

"It was Paul, well, I mean, the other Paul's idea. He said to call him once you found out."

Christine turned to pick up her watch from the bedside table and screamed at it. "Maya, call Paul, NOW!"

He picked up almost immediately.

"How could you do this to me, Paul?" Christine said, quavering notes of thinly suppressed anger in her voice.

"I am so sorry to upset you, Chrissie. I know this is awkward. I mean, beyond awkward."

She groaned. "Awkward? Is that the word you think describes this? This was an attempted rape, you fuck."

Paul was quiet for a moment. "I thought if a Transfer could pass *this* test, it could pass *any* test. I didn't think it would do any harm, I mean, other than the surprise."

"No harm done! Fuck you, Paul. I am not a fucking lab rat. I've been played. By you, of all people! I hate you so much right now!" Suddenly she bent forward as dry heaves clutched her, her soul leeching pain.

Paul stayed silent, as if waiting under a bridge for the dark cloud to pass. Paul Prime stood ten feet away, inert. Christine couldn't stop herself from stealing repeated short glances at him.

After a few minutes, when her body recovered a semblance of normality, Paul said, "I understand, but please, Chrissie, as I see it, it was for a good cause. Now even you must be convinced that Transfers are excellent. You really didn't know?"

She narrowed her eyes, even though he couldn't see her through the voice-only connection. "I mean, not until I saw the barcode."

He hesitated, then said, "Actually, I'm nearby in a kind of nuclear bunker in the mountain behind the house. I will be the one at the meeting tomorrow, not my copy. The Transfer will hide in the bedroom closet, and I'll deactivate him. We can talk more in person. But please, Chrissie, please don't tell anyone. Bart knows, but not Jeremy or Koharu."

"I…okay," she murmured distantly, then disconnected the call and stormed out of the room without another word. She popped two sleeping pills and pushed them down with a full glass of wine.

CHAPTER 9

The next morning, Christine awoke to the coo-woo-woo-woo of a turtle dove. She had slept poorly, lost in an ocean of thoughts with no shore in sight. After she showered and dressed, she headed for the kitchen. Coffee might help, even if that meant facing Paul. As she passed his bedroom door, she thought for a second of checking the closet to see Paul Prime, but the thought gave her the creeps.

Bart and Jeremy were already in the kitchen when she arrived. Christine nodded to them and set to making herself a coffee.

Koharu walked in a few minutes later, poured themselves a large mug of hot water, and added a tea bag. They'd all settled into a comfortable silence when Paul came up from the basement.

"Oh, I thought you were upstairs," Bart said, pouring himself a large mug of coffee. "What were you doing down there?"

"Picking wine for later." Paul glanced at Christine.

She narrowed her eyes, trying to suppress the wave of nausea that flooded her at the sight of him.

"Everything okay, Christine?" Jeremy asked.

"Yes, thanks, Jeremy." She avoided his eyes. "I guess seeing that…mouse disturbed me more than I realized." She walked with her coffee mug behind Paul, trying to get a look at his neck, but he was wearing a turtleneck. Bart caught her eye and gave her a *don't do that!* look.

"There's cereal and toast. Anyone want eggs?" Paul asked, deftly changing the subject.

"I would," Bart said. "French Canadian style?"

"Hmm, decadent," Paul said, smiling as he pulled eggs and butter and syrup from the fridge. "Sure, why not?"

Paul and Christine had once been to a sugar shack near Montreal and had a wonderful meal, all maple based. Scrambled eggs cooked in a half-inch of maple syrup, served on thick white toast with loads of butter. They'd tried to repeat the experience at home and failed a few times until they found the right syrup temperature to cook the eggs. The memory-filled her with an unsettling mix of fondness and revulsion.

Paul slipped a small piece of paper in front of her before turning to the stove. It read, *All dear dreams come real now, and ready the blessed gifts of love.*

Christine could feel her blood boil. As if a Pushkin verse could begin to unwind the hurt in her soul? She was much too acutely angry for that, though he seemed to be completely ignoring the waves of hostility coming off her as he cooked and served their breakfast.

"Wow, these are good," Jeremy said after taking his first bite. "It would never have occurred to me to cook eggs in maple syrup."

"The trick is do it slow," Paul said, looking at Christine as he helped himself to a huge portion.

When they were done eating, they each grabbed their mugs, ready to head downstairs to the vault.

Paul tapped on Bart's shoulder. "A word?"

They walked out onto the patio for a few minutes, and when they returned, Bart announced that he had to call someone at the office.

"Let's reconvene at eleven, all right?" Paul suggested

Jeremy frowned and seemed about to protest when Koharu said, "Actually I'd love to have a bit more time to catch up on the latest reports from HQ."

They all got up, and Paul motioned to Christine to follow him upstairs.

***

Paul closed the bedroom door behind them, and Christine sat on a dark blue velour armchair in the corner, under a reading lamp. She grabbed the switch and then changed her mind, leaving it off. Paul remained standing.

"Look, Christine, I know you're very upset."

She didn't have the fortitude to answer, so she just stared at him with empty eyes.

"Will you just give me a chance to explain?"

She crossed her arms protectively in front of herself. "Show me your neck."

Paul looked nonplussed for a second, but he complied.

When Christine saw there was no barcode, her body relaxed a tiny bit. Paul walked over to a pine armoire in the corner with chiseled doors. He opened it. Christine gasped, even though she'd known what she would see. Inside was, well, another Paul.

"This is Paul Prime. You…met him already."

"I didn't just *meet* him, Paul. He, or *it*, was planning to have sex with me… that…thing," she spat, the words tasting like poison in her mouth.

"What I need you to understand, Chrissie, is that this…" he gestured towards the armoire, "is me, but a better me."

"Better?" Christine's laugh gushed with mockery.

"Yes, Chrissie." He sat down on the corner of the unmade bed. "You know, I've been taking maximum doses of so many pills. Bupropion, sertraline, and too many others to count. I'm barely holding it together. There are many days when I just cannot work. I spoke to Bart, and we agreed that he would try to work with Paul Prime on the project on days where I can't. And it's worked out well. Better than we had hoped. Paul Prime is never tired. He thinks like me, has the same memories. He *is* me."

Christine rolled her eyes. She was ransacking her mind, trying to make some sort of sense of it all. On her right shoulder, an angel voice was saying that Paul wouldn't have done that to her if didn't mean so much to him. But then there was Rachel's voice on the other shoulder: *Fuck the bastard.*

Paul swallowed. He was making his best puppy face, like a dog who just ate the pie that took hours to bake right before guests arrive but felt really guilty about it. "I have a big ask, Chrissie, but you can see how important this project is for, like, the future. Everything."

Christine didn't answer, and Paul moved over to look out the window, giving her time to process.

Finally, she said, "How can I tell you apart?"

Paul startled, and his sullen face softened when he turned to look at her. "Not always, and that's the whole point. But there's the barcode, as you know. And Transfers don't eat. They can drink fluids and even have something like taste buds."

Tension she hadn't realized she'd been holding flooded out of her body. "So it is you. You wolfed down two plates of eggs saturated with maple syrup."

"Yes, a last meal, for a while at least. Not a bad one, I might add."

Christine sat up, shifting instantly from anger to worry. "What do you mean, a last meal?"

"I really cannot function normally, Chrissie. Not until there's a new class of drugs. So, I've made arrangements to be, you know, frozen."

"Cryotherapy?"

"I think you mean cryonics, but yes, that's basically it."

Christine interlaced her fingers and lowered her head to look at her hands. As if they contained an answer. "Where?"

"Colorado. Best in the nation."

She fell silent again, mind spinning with the implications of what he was saying. "When were you thinking of doing that?"

"I plan to leave very discreetly when the meeting resumes today and head in that direction. Short flight."

"Today?!" Christine stood and walked towards him without conscious thought. She was still so angry, but other emotions were jostling in her heart. "Oh, Paul."

Christine opened her arms, and Paul stepped into the hug. They stayed like this for a while, and then Paul took a small step back, grabbed Christine softly by the shoulders and looked straight in her eyes.

"I hope you will understand, Chrissie. I love you. I always will. But I can't really love right now. Not with a broken mind. I want to leave you with—he turned his chin towards the armoire—him. I'm not asking you to love him, or even like him. But I'd like you to help him with the project just like you would help me. It is *me* you're helping by helping him. When I come out of deep freeze, I will do my best to repay you, if there's ever a way I can do it."

She looked out the window, then turned her head back towards him. "I'll think about it. Right now, this is all just a big jumble in my head."

"I understand. Do you want me to ask Bart to postpone the meeting? As I told you, he knows, but the others don't."

"Why haven't you told Jeremy and Koharu? You don't trust them?"

"Bart and I have discussed it many times. We're just not sure how they would react, and we cannot take such a big risk at this stage in the project."

"I see." Christine walked up to the window. "I'm going to go back to my room for a while. I need to get my thoughts in some sort of order."

"Sure, Chrissie. Should we postpone then?"

She looked at her watch. "No, I'll be fine."

Christine started walking towards the door, but then the realization slammed into her that this may be the last time she ever saw him. She turned around and ran to him, throwing her arms around his neck. They hugged for several long minutes.

When they parted, he gave her a soft smile. "Goodbye for now, Chrissie."

Unable to answer, she simply nodded and left. As soon as she got to her own room, she started crying, softly at first, and then her whole body started shaking. She mechanically opened her purse, planning to pop two of the antianxiety pills she always carried with her. There was a white, card-size envelope in the purse. Handwritten. Her hand shook as she opened it and pulled out the cream-colored card.

Goodbye, my friend, goodbye,
You touched my heart, my dear.
Our preordained separation
Signifies we will be reunited one day.
Goodbye, my friend, without a handshake, without a word.

Don't be sad and don't frown.
Dying is nothing new in this life,
But it's certainly also not new to live.

Yesenin's last poem, just before his suicide. She fell on the bed as her tears quickly turned into deeper, pregnant sobs. Eventually, she managed to regain enough wherewithal to pop three pills and take a sip from the glass of stale water on the small, round table next to her, under an old lamp with its leather lampshade discolored by time atop a brass base. She waited a few minutes and felt the pills kick in, allowing her to regain her composure, part of it anyway. It would have to do.

***

They reconvened in the basement at eleven, as planned. Bart looked at Christine when she entered, a question in his eyes. She nodded discreetly, and Bart nodded back.

Then he turned his gaze to the whole group. "Let's pick up where we left things yesterday. We were discussing whether having a human personality transferred into a robot might change the ethics of the situation for us. Christine, do you want to get us going?"

"Sure," she said, trying to avoid the imposter Paul's gaze. She needed to get back in the groove of this discussion and push what had happened out of her mind, at least for now. "I think there are a number of questions that existing codes of ethics would not provide answers to. The first one is, how do you reconcile the fact that there will be two almost identical persons, those you called A and…A-Prime yesterday," she said, trying even harder not to look at Paul's doppelganger. "You would have two persons, but as a matter of law, would they both be in control of the same thing? Would A and A's copy have the same house, bank account, family? If so, you have an almost impossible situation. Besides, unless they shared all future experiences, they would grow to be different over time."

"I see the problem," Paul's replacement said. "Before we continue, could I suggest that instead of calling the robots copies or clones, we call them simply 'Transfers'? After all, that is what they are."

No one spoke, and Christine furtively looked from Bart to Koharu to Jeremy, all of whom seemed mildly confused by the question.

With a warning glance at "Paul," Bart said, "Sure, for now. Maybe we revisit later. But that's not the real issue we were talking about. Christine, what do you think? What if we had people and their *Transfer*," he said making air quotes, "both, you know, going around doing stuff?"

Christine kept her attention fixed on her coffee mug. "I think it could lead to serious liability and other issues," she said, in a voice barely loud enough for the others to hear.

"No kidding!" Jeremy, who was sitting next to her, agreed. "I mean, other people won't be able to tell the difference. They won't know who, or what, they're dealing with."

"I can see a solution," Paul's copy said.

"Namely?" Bart asked.

"Well, it might sound kind of drastic. But the best way to solve the problem is to avoid it entirely."

"Now I'm afraid of what you're going to say next," Christine said sardonically.

"Paul" did not seem to pick up on her tone. "I think the answer is to only allow personality transfers after the person is dead. After all, the data won't die with the person, right?"

"Allochrony, in other words," Koharu said.

"Pardon?" Bart asked, looking at her. "Allo what?"

"That's what it's called in biology. When two entities cannot exist at the same time."

After a brief silence, Paul's copy swiveled his white leather chair towards Christine, "Don't the heirs of a person inherit the rights to the data that person had when they died?"

Bart was looking at her eagerly, awaiting her answer, and much as she wanted to be passive aggressive, Christine knew she would not be too successful with that approach. "Well, to some extent," she said carefully. "There's no obligation to delete anything when the person dies. But presumably, each person who wants to transfer would have signed a contract, and that contract would bind the heirs."

"And *presumably* the transferred person can be the heir?" Bart said with a smile, repeating Christine's word slowly.

She nodded. She could see from the corner of her eye that "Paul" was smiling now, too.

"Hmm, I guess that solves that problem," Bart said. He took a sip from his coffee and looked around the table. After a short silence, he said, "This means our market is dead people only. Right, Paul?"

"Well, that," he said, "or people willing to die."

Koharu shrunk back in their seat, gaze darting anxiously around the table before going back to the robot version of Paul. "What do you mean?" they asked softly.

"I mean that we could offer the service, but people would have to agree to die to be transferred."

"You mean assisted suicide?" Koharu still looked like a kid who'd just been told their parents were killed in a car crash.

"Yes, something like that," Paul's copy said matter-of-factly.

Koharu's eyes hardened. "But isn't that only legal for people with terminal diseases, and then only in a few places?"

"I know. We may have to move the execution of the plan..." Paul's Transfer began.

Bad choice of words there, Christine thought.

Perhaps realizing this, he added, "I meant, we could say that the service is offered in certain jurisdictions only. I think the Netherlands and Switzerland have fairly open policies." He kept his eyes on Christine, who now had her face composed as if for a visit to a funeral home.

"I would have to check," she said, "but in theory, it is possible."

Bart pushed his chair back. "I had never thought of it that way, but in fact, it does make sense. And yes, you're right about the Netherlands." He took a long sip of coffee. "Up to now we considered that the Transfer, as you want to call it, Paul, or A-Prime, could survive after A's death, unless we programmed it to self-destruct upon A's death. That is the problem we had been trying to solve, but now I see it was the wrong one. What you are saying is to make sure that there is only one person at any given time. A dies, A-Prime takes over. This is a big change to the way we will need to sell the service. Instead of selling people a copy of themselves, we would be selling something like...immortality."

"Yup," Paul Prime said. "Not a bad product. Transfers would no longer be for the sake of having a body double, but a new you, a 'forever' you."

There was a long silence.

"Couldn't we freeze people at least?" Koharu asked. "It seems much less radical. Cryonics has made so much progress in the last ten years."

"A *definite* possibility," Paul Prime said, giving Koharu a strange smile that made Christine's skin crawl.

No one spoke for more long moments.

"The dream of immortality, finally a reality?" Bart said, breaking the silence. "So, if I understand the proposal, this mean that we would offer transfers and assisted suicide as a package. What do we all think about this?"

"And cryonics, too," Koharu added.

"I think as a selling point, some version of immortality is worth a lot more than a new robot friend," Jeremy said. "As a business, it makes a lot more sense. Fewer customers, but a much higher margin."

"Yes, I can see that," Bart said. "But we all agreed we weren't in this for the money, right?"

"Paul" was looking at Jeremy. "What do you mean when you say 'some version,' Jeremy?"

"Well, the person actually dies, right? Their personality survives, but only for others. I mean, if you're dying and you can transfer so you can continue to raise your kids, love your spouse, help your community, and do whatever else you did, then it's a version of immortality. But the person is dead, so it's not *real* immortality."

"I see. Christine, what do you think?" Paul Prime asked.

"I think that's right."

"Paul" frowned. He looked at his hands and said, "Your new 'you' can live without fear of death or disease. I mean, it is a pretty great version of immortality."

"Well," Christine said, "that may be. And maybe it answers the first question, but it raises another."

"Which is?"

"I suspect only the ultra-rich will be able to afford this service. That means a planet full of humans living normal lifespans and ultra-rich human robots continuing to live forever, probably amassing knowledge, fortune, and power and ultimately controlling everything."

"You're right," Koharu said, looking at Christine. "We cannot offer free transfers to every person on the planet. And with the planet already at full capacity, we would have to consider how this affects ecosystems."

"Fewer humans means a better planet," Paul Prime countered. "Humans are causing climate change. The sooner they die, the better!"

They all stared at him with eyes the size of half-dollars.

"Transfers don't need income," he continued, unfazed. "They can exist without food or even shelter. Maybe the ecosystem catastrophe can be avoided once many, many people have been replaced. In fact, if everyone was transferred, no one would need to eat."

Another awkward silence. They all grabbed their mugs more tightly. Then something occurred to Christine.

"Once you make this technology public, I suspect the stock of companies that grow and sell food will drop."

"Not our problem," Paul Prime said.

"A major economic downturn wouldn't help anyone," Christine said, trying to refocus him.

"Paul" frowned. "True, but technological progress must continue. Didn't someone say that before, something about inevitable creative destruction?"

"Yes, Schumpeter. An economist," Bart said. "Mandatory reading in business school. But Christine has a point. We don't want to be the cause of a global financial collapse."

"As I see it," Christine said with a small but unmistakable victorious grin, "there may be a solution to that problem."

"Okay!" Bart said. "Let's hear it."

"It's probably my university bias, but what about a needs-blind business model?"

"Needs-blind?" Paul Prime asked. "As in, pay what you can?"

"Something like that."

"That might work," Bart said. "But as Koharu said, we still cannot service 8.5 billion people. We just cannot scale up to produce that many units."

"That's another hard question, possibly the hardest," Christine agreed.

"Well, before we get to that one, may I suggest a coffee break?" Bart said. "I think we could all use one. I know I do."

As the others headed for the kitchen, Paul Prime stood and walked over to Christine's chair. "Thank you for all you did during the meeting."

She was mildly nauseated, as if the emotions of the previous hours were stuck in her throat. "You know, *Paul*, I almost feel like I'm in some sort of Jonestown cult. You're so weird. And this trick you two played on me. Fuck."

"Jonestown?" Paul Prime looked affronted. "Nothing whatsoever to do with that. Is it suicide if it leads to immortality?"

"Yes, *Paul*, it *is* suicide. The *personality* continues to live, but the *person* is dead. And what about this transfer process? Are you sure all of the person will survive?"

"What do you mean?" He lowered his voice to a whisper. "I thought I convinced you. We were together for some time before you noticed anything."

"I mean, okay, yes, it was a very good copy of, well, you, but it was for a couple of hours. How can you be sure the whole person can be transferred? Whatever the soul is, you cannot transfer it—all 21.3 grams of it."

"Ah yes, the old Duncan MacDougall theory." Paul Prime chuckled. "Look, we're transferring data. I mean, it's like we're transferring the *empirics of the soul* so that the Transfer will act like the human it copies. Or maybe I should say 'replaces,' now that we've decided that."

"Yes, but how will future events affect the Transfer's behavior? You can trick people into believing the Transfer is the same person, but for how long?" When he didn't answer, she stood. "I need a bio break."

When she came out of the powder room, Paul Prime and Bart were gone and Jeremy and Koharu were both typing on their tablets. She made her way upstairs. As she approached the kitchen, she heard Bart and Paul Prime talking.

"You had an excursion in mind. Right, Paul?" Bart turned as she entered the room. "Ah, Christine. Maybe I should leave you two?"

Christine held up a hand. "No need, Bart. I want both of you to hear this. I've decided I will play along, but I am doing it for Paul's sake. So you..." She pointed at Paul Prime. "Please don't ever try to act with me like you're the real thing."

"But, Chrissie..."

"And call me Christine." She had iron shutters on her face.

Bart took a step towards her. "If he does that, it will be obvious something's amiss, don't you think?"

Christine hesitated. She looked out at the mountains on the horizon, looking for an anchor. "I guess," she finally agreed. "But only in public." Paul nodded.

Bart put his finger in front of his lips and pointed towards the stairs. A few seconds later, Koharu and Jeremy entered the kitchen.

Jeremy frowned. "What's up? I thought we were just taking a few minutes."

"Well, it's almost lunch time," Bart said. He turned towards Paul Prime, who flashed his most charming smile.

"I've arranged a quick trip to Yellowstone Park. I have jetpacks ready for everyone, so we can be there in fifteen minutes."

"Oh, I've always wanted to see Yellowstone!" Koharu said, instantly reenergized.

They followed Paul Prime behind the house to what used to be a four-car garage. Now it housed only an old grey Maserati coupe covered with a green tarp and five new-generation jetpacks lined up against the wall.

"These are like the old ones you've all used before, just easier to operate and much faster. A hundred and fifteen miles per hour." He handed out gloves and special helmets. "The AI is also a lot better. You can give it a destination using voice commands or use the helmet camera to scan coordinates on a map."

Once they were suited up, Paul Prime spoke through his helmet mike. "Everyone ready?"

The others all gave a thumbs-up.

"Aim your cameras here." He pointed to the coordinates on his tablet, and they all scanned the data and then exited the garage to spread out on the lawn.

Paul Prime said, "Go!" and his jetpack beeped and slowly started to climb. The others followed.

As they ascended, Koharu said into their microphone, "What are the mountains there on the left, with the snow on top?"

"The Grand Tetons," Paul Prime replied.

Christine remembered how she had laughed the first time she'd seen those. *Grand Tetons, you must be kidding! Why not Big Boobs?* She'd later learned that that was exactly what the French explorers who'd "discovered" them intended.

They flew around the Tetons and their blurry reflection in Jackson Lake, painting a majestic landscape that could make anyone a tree hugger, at least for an instant. They passed near Huckleberry Mountain, where a few ochre-colored autumn leaves clung to their branches, as if afraid to be swallowed by the earth beneath them. Then Mount Sheridan, a heap of dough that the Earth had forgotten to knead. On the other side of Mount Sheridan, the majesty of Yellowstone Lake rose in the distance. A few minutes later, the jetpacks automatically started their descent and crossed the Lower Falls, its frothy waters ensconced in fir trees and golden rock faces. They landed in a meadow overlooking the falls, where a large table with a red and white tablecloth and a warm lunch awaited the group, brought by robots from the mansion.

"That was amazing," Koharu said, taking off their helmet. "It's even more beautiful than I imagined." Their face was lit up like a kid on Christmas morning.

For a while, they all ate their chicken cacciatore in silence, in awe of nature's grandeur around them, while Paul Prime lay on his back looking at the sky, a glass of San Pellegrino next to him. When she finished eating, Christine did the same, but with enough distance between them to discourage conversation without arousing suspicion.

She was not so much gazing at the sky as she was grazing it with her mind. The deep blue immensity in which her mind was traveling freely made her realize that the questions they were discussing were not just business matters. They were at the core of what it means to be human. It wasn't the arms, legs, or body, because someone with artificial limbs would still be considered human. So, it had to be the mind, and wasn't that what Eidyia was transferring? Was the mind the same thing as someone's personality? The old idea that the soul rests in that pump in the middle of our chests had to be wrong. The soul, whatever it was, had to be in the mind. But even an enhanced mind was human, so what then?

Koharu's voice interrupted her thoughts. "Seeing a place like this, I think I know what Jeremy meant now when he talked about making it better."

"Hmm," Bart said. "What is that?"

"Well, there are people who are doing what they can to make the planet a better place, and those who are doing the opposite. It's like, I don't know, short-term greed and closed-mindedness on one side, and a better future for all on the other."

"Not a bad yardstick," Bart said. "So, we only transfer ecologists and humanists?"

"I didn't mean it that way. Just maybe people with some sort of conscience?"

"Conscience? You mean people who think about the consequences of their actions on others?" Paul Prime asked.

"Yeah, I guess that's what I meant."

"What do you think, Christine?" Bart asked.

The question shocked her out of the torrents in which her thoughts were trying to stay afloat. She tried to get her law professor gears back on track. "Well, what makes a person good is an old question. In law school, we study philosophers all the way back to the ancient Greeks, and even before them, who have tried to solve that equation. There is no uniform answer. Some say that we are all driven by our passions and trying to do what's best for each of us, and that law's job is to put limits on what each person can do and to secure property rights because a system in which everyone is entitled to everything cannot work. This is kind of what *The Leviathan* is about."

"*The Leviathan?*" Koharu asked.

"A book by Thomas Hobbes, a seventeenth-century English philosopher. The premise is that there are finite resources, and you cannot take or damage what belongs to others. Otherwise, you would destroy agriculture and industry. But with nature, like this park, or the environment, it's a bit different."

"How so?" Bart asked.

"If my business destroys your land," Christine explained, "you can go to court and get compensation. There is a cost imposed on me by the state, in other words. This is the role that Hobbes thought the state should play. If, instead, I pollute the planet, but not *someone's* land, it's very difficult to hold me accountable because it's no one's land, or everyone's."

She took a sip of her water, looking pensively at the horizon. Why did people do that? Was the ineffable lurking in what looks to us as infinity? Wilkie Collins's words came back to her:

Sympathies that lie too deep for words, too deep almost for thoughts, are touched, at such times, by other charms than those which the senses feel and which the resources of expression can realize.

She glanced back at the others and found them all giving her peculiar looks. *Better get back to the here and now.*

"The law is much better at protecting the individual than the collective," she said. "As a result, there is often no cost to pollute. I can endanger the health of everyone on the planet, and all future generations, and if that increases my profits, my incentive is to do it as much as possible. That's what the oil and gas companies did when the climate change debate started to pick up steam in the late 2010s. They funded junk science studies and launched PR campaigns showing a few wind turbines here and there. This was hypocrisy of the highest order. They even paid for media campaigns to discredit a teenage activist who ended up being the first recipient of the new Nobel Planet Prize a few years ago."

"I think I get it." Koharu's brow wrinkled in concentration. "I wonder if an algorithm can help measure this."

"I think it can. In fact, I *know* it can," Jeremy responded.

Everyone's attention turned to him. He'd been silent since their landing on the meadow.

"You *know*?" Koharu gave him a funny look.

"Yes. Eidyia decided to tweak its search algorithms years ago so that climate change deniers would be ranked lower, and all their information accompanied by disclaimers. Well, that means AI could identify people and companies that lie or are detrimental to the future of the planet in other ways. Racists, for example, although it's so systemic that we'd all fail that test. But overt racists, maybe. All we would have to do is ask the algorithm to apply it to each individual who wants to transfer."

"But wait," Christine said. "Do you want the environment and racism to be the only concerns here?"

"You have a point," Bart said. "What do you suggest?"

"Honestly, I'm not sure. I think you could identify other factors. Generosity maybe. Kindness. Couldn't you program an algorithm to help you decide what the criteria should be?"

"It might be worth trying," Bart said.

They fell silent, each lost in their own thoughts, until Jeremy abruptly stood. "Anyone up for a hike?"

The energy of the group shifted immediately, and Christine breathed a sigh of relief.

"I am!" Koharu said.

"Me too." Christine's brain needed a pit stop, and she'd been sitting for most of the morning.

"Let's go, then," Bart said. "Paul, any suggestions?"

Paul Prime lifted his head and looked at them. "Hmm, yeah, sure. The South Rim trail is right over there." He pointed to the trees behind them.

"Not in the mood for a hike, Paul?" Christine asked, trying not to choke on the word.

"Sorry, Chrissie. My mind is just fulltime on this stuff, and I find it hard to shift back into neutral, so to speak."

"I see." She eyed him suspiciously as he rose to join them. How could the others not notice how strange he was acting?

They followed Jeremy up a small hill in the direction "Paul" had indicated and then down a small path going through a tuft of evergreens. They walked in silence, marveling at the gorgeous scenery around them, Paul Prime bringing up the rear.

When Jeremy slowed down a bit, Christine caught up to him.

"I'm glad you brought up the issue of racism," she told him. "It's an important factor to consider if we're talking about creating an algorithm for 'goodness.'"

"Yeah. There's a lot of bias in the system that we need to factor in. And it's kind of personal for me. My brother was caught with a small amount of pot, three grams, I think, before it was legal, and…"

"But it's been legal in California for so long. I thought YOU grew up in LA?"

"Yeah, but my brother was in Texas for college. A full scholarship in Austin."

"Ah, I see. Fuck."

"Yeah. So, long story short, he lost his scholarship, and then got a year in jail and lost everything else. The white boys he was with got a two-week suspension."

"Racism flows through the veins if this country like a putrid drug, as one of my law professors used to say."

"Exactly, and so if we have a chance to detox people from their racist dipshittery during transfers, I think we should jump on it. You know, it had never occurred to me before. But now I love the idea."

"I can see why! But like you said, in a way, we are all potentially racists. But few people are consciously racist, and even fewer would admit to it."

"That is *exactly* the point, Christine. I agree. Data doesn't lie. It…" He stopped as they came out of the trees and saw the falls carving their path through the rocks and extending their silky web far into the valley ahead.

"Man, it's so beautiful!" Bart said as he caught up.

They were catching their breath when Paul came out of the trees looking fresh as ever.

Bart leaned into Paul so that no one else could hear, except Christine who was just a few feet away. "Didn't break a sweat, did you?"

Paul just looked ahead.

When they'd all gotten their fill of the view, he asked, "Should we get back to Jackson?"

They hiked back down to the meadow, which the robots had already cleaned up, and zoomed back to the mansion with the jetpacks.

***

Back at the conference table in the basement, they all sat down with fresh mugs of coffee, except for Koharu, who had their usual green tea. When everyone was settled in, Bart turned to Christine with a smile.

"I guess it's your turn again, Christine. What's your final question?"

She pursed her lips and put her mug down on the table. "It's also a hard one, I'm afraid."

"It's not like the rest of them were easy," Bart said, looking a world more relaxed than before the break.

Christine spread her hands in front of her. "The question kind of came up this morning, but to put it in a nutshell, when people transfer, do you make them transfer as they are, or do you let them improve?"

Jeremy smiled.

"Please continue," Bart said. "I think I see where this is going, but I need to hear more."

"I might call it the 'tel quel' question."

"Tell what?" Jeremy asked, furrowing his brow.

"Tel quel. It's a French expression. It means something like 'as is.'"

Jeremy nodded in understanding.

"We also have something like it in Japanese," Koharu interjected. "We say sonomama."

"Sounds like something you'd want to eat!" Bart joked halfheartedly.

Koharu gave him the look.

"What I mean is that we absolutely need to avoid falling into the trap of ableism."

"Ableism?", said Bart, raising his brow.

"Yes. Say someone who wants to transfer has a physical issue. Maybe the person is paraplegic. What you plan to make is a copy, right, so the new body is an *exact* replica?" or "Would the copy share that disability???"

"Why wouldn't we allow people to transfer without the disability, to be cured you know," Bart asked, looking surprised. Christine smiled benevolently. She had heard this too many times.

"Well, I am not saying people should not be allowed, but it should most definitely not be presented as a cure or something like it. As far as I know, many people don't feel there's anything they need to be "cured of." Look, my dad is autistic, and neither he nor I ever understood why that's a pathology. I always understood it as a relationship between the mind and the outside world that differs from the dominant norm. Not everyone wants a neurotypical brain. We should want the exact opposite in fact, to accept and accommodate neurodivergent selves. So it's really tricky when you're talking about 'cures'. Now, cures to fatal illness like cancer or heart disease or organ failure, I'd think people would pretty universally agree that that would be a good thing. But many disabilities are trickier." "Hmm, I never thought of it that way," Bart said.

"I still don't see a problem," Jeremy added, tilting his head. "People can choose."

"Well, it's probably not that simple," Christine said, looking at Jeremy. "A physical constraint influences how a person functions and behaves. And not just the obvious physical limitations, but the psychological impact, positive or negative. What I mean is that the person would not only be changed physically but also mentally. Maybe not in a bad way, of course, but if you remove that, it's no longer just a transfer"

"What is the answer to your riddle, then?" Bart asked.

Christine squeezed her mug with both hands, as if trying to discover a simple answer in the magic cauldron. She felt everyone's eyes on her. "I think you could argue that if it's just like medical progress, it's completely defensible. In terms of disabilities, I don't think anyone in this room can make that kind of decision. We would need to get people with different abilities to tell us." So basically I guess Jeremy is right, they can choose, but how we offer and frame that choice is crucially important."

"That makes sense," Bart said.

Jeremy nodded.

"So, it wasn't so hard after all, that last question," Bart said. He got up to make another cup of espresso. "How's the macchiato, Christine?"

"Oh, hmm, it was excellent, thanks," she answered distractedly. Paul Prime had been unusually silent, and she found it odd that neither Jeremy nor Koharu noticed what struck her as a different demeanor. "But that was the easy part in a way."

"Damn," Bart said, still smiling, but the lines of his forehead let more than a hint of worry show on his face.

Before Christine could continue, Paul Prime said, "I think we've progressed quite a bit today. We've earned a glass of Elba!"

"Elba?" Koharu asked. "Isn't that the name of the island where Napoleon was imprisoned?"

Christine frowned at Paul Prime in confusion. She was about to unroll *the big question*, and his contribution was to change the subject? What the hell?

"You know your European history!" "Paul" said, clearly relieved at the shift in subject. "More like exiled, but yes, they make splendid wine. Napoleon apparently had more than a few glasses of it during the months he spent there."

Christine looked at Bart, whose face had softened in a heartbeat. Was it the relief of postponing a hard discussion? Or had he understood something different from Paul's interruption?

"Can we speak about wine later?" Jeremy asked, visibly irritated. "Christine, please, can you continue what you were saying?"

Koharu nodded gently. Paul Prime, who had just gotten up from his chair, sat back down, clearly annoyed. Christine looked at Bart, whose face was now closed, but he nodded.

"When you change people, you change society," she said. "It's really social engineering on a massive scale. But from the inside."

A flicker of confusion crossed Bart's face. "From the inside? You lost me there."

"I mean that social engineering is usually something imposed by a government, like what the Communist regimes did with central planning. It changes the structures of society and the institutions and uses propaganda, and in time, it changes people's behavior." Koharu was looking at her intently. "History tells us, of course, that central planning always fails, often due to the progressive emergence of corruption, but sometimes simply because it's impossible to plan the future on such a scale. For humans, anyway. But social engineering can be a lot of other things. Advertising is described by at least some people as a form of social engineering with a limited goal of manipulating consumers into making purchasing decisions."

"Social engineering isn't necessarily a bad thing, then?" Koharu asked.

"Well, I'm not sure that what people have said about social engineering applies to this situation. We are talking about a totally different kind of social engineering. Instead of changing the structures and the institutions from the outside to influence behavior, you change people, and then they change the structures and institutions. That's what I mean by 'from the inside.'"

"But how do you get that right?" Bart asked, moving forward in his seat.

Suddenly, Jeremy stood up, moved behind his chair, and put both hands on the back of it, leaning forward. That got people's attention. "Look, the technology is hard, but the rest is easy. We said we weren't in this for the money. Fine. But we have a chance to eliminate or at least reduce bigotry, so what's the big question? Either we say no to bigots who want to transfer, or we filter bigotry out. What can be bad about that?"

Everyone looked at Jeremy. Christine nodded just enough, she hoped, for Jeremy to notice.

Bart took a sip of his espresso. "How would you do it, Jeremy? You just said the technology is hard."

"You can use personality profiles, I guess."

Bart turned to Koharu. "Koharu, from your perspective, is this doable?"

"Theoretically, we could use psychology, evolutionary biology, and other recent advances to identify racist behavior."

"Exactly!" Jeremy said. "That's what I'm saying. Racism is an empirical thing. It's not whether you believe you're a racist or not. It's whether your *behavior* is racist." He paused. "And our business is data. We have the data." His face brightened with excitement.

"Geez," Bart said. "This project is way more far-reaching than I ever imagined!" Jeremy sat back down, smiling.

After a moment, Christine said, "Maybe I do have a suggestion after all."

"Good," Bart said, "because frankly I'm swimming in my own thoughts here, and I no longer know which way the ladder is."

"The solution might be to avoid this problem like we've talked about avoiding others."

"How would you do that?" Bart asked.

"I mean, let people who transfer solve physical impediments that medical science might theoretically solve one day, but don't touch psychological profiles. Especially if you're going to filter out applicants, you may achieve social engineering effects but without the Frankenstein risks."

"The Frankenstein risks? Like the monster?" Koharu said.

"Frankenstein is the name of the scientist who created the monster actually, but the point is that the monster is intelligent and has self-awareness. It even calls itself a fallen angel, a Lucifer of sorts. I assume you want to have persons after the transfer, not some sort of creature based on a human personality but changed. I mean, if you tamper with personality, who knows what might happen."

Jeremy looked at her, visibly unhappy.

"Okay," Bart said. "I don't think we get anywhere today or even tomorrow if we continue. We all need to think on this a bit more."

Paul Prime gave him a dim smile. "Who wants to try that Napoleon wine now?"

***

They moved to the garden. The temperature had dropped, and rain was approaching. But around the fire pit, with glasses of Elba and old Italian cheeses cut into small cubes, it was pleasant to be outside. Everyone was silent for several long minutes, until Paul realized they were short one person and went to search for Christine.

He climbed the stairs and opened the bedroom door to find her standing at the window, looking at the mountains.

"What's wrong? I mean, besides the obvious."

"I don't know exactly. I mean, here we are discussing what may be the biggest step towards eliminating racism, sexism, classism, and all the other false divides that fracture human society, and I'm…how can I put it? Of two minds about it. What do *you* think?" She paused. "What did Paul think? You didn't say much."

"I think you're looking at it from the wrong end. That's why I tried to call a break, so we could talk off-line. But Jeremy prevented that. The good thing is, we didn't get to a conclusion today." He came closer, stopping when she tensed, and looked out the window with her.

"The wrong end? I have no clue what that means." She turned towards him, and he looked in her eyes.

"I mean, persons *will* change after the transfer. You don't need to change them before or even during the transfer."

"You mean, they will continue to evolve?"

"That, but also they will all be connected in some way."

"Connected?"

"Yes, the Grid is not merely a power source," he said, referring to the network that powered all Eidyia robots and many other AI devices. It had quantum computing capability and sometimes helped robots process data. "It can … I guess you could say it can slowly reprogram people, in a way."

"Wow. That changes *a lot* of things."

"It changes *everything*, Christine."

CHAPTER 10

Christine had left Jackson in a state of great mental confusion. Back home and trying to prepare for class a week later, she wasn't sure how to maintain a wall between what she now knew about the Transfer Project and her teaching, but her answer to Bart's question about what makes a human human was constantly gnawing at her brain. She couldn't say anything to her students about Eidyia's project, but maybe she could get fresh ideas from them. She now had an idea for how to define a human as a matter of law and planned to relentlessly steer the discussion on that question.

Perched on the table in front as usual, she said, "As you know, today's class is about robot rights. This is a difficult topic, and I hope you have all read the various texts I sent you. Opinions range from no rights at all, like the chair you're sitting on has no rights, to a set of limited rights like how we recognize a right for animals not to be mistreated or have unnecessary harm inflicted upon them. Then there are those who say it's time to give robots rights that parallel those of humans, maybe even the exact same rights. So, let's begin. Who wants to get the ball rolling?"

Today, there was a volunteer.

"I don't think robots should have any rights, Professor," Mira said. "They are things, and things have no rights. They are objects. *We* have rights *over* them. Animals may have some rights because they are alive, but robots are not."

"Thank you, Mira-san. Looks like you are well prepared for class!" Christine said, but Mira's smile evinced doubt. "Great way to start the discussion. Who agrees or disagrees with Mira?"

"I think Mira's kinda right about robots being things," Charles said, "but animals are also things. We can buy and sell them. I think we give them rights because they can feel things, like pain. Robots don't feel anything."

"Interesting," Christine said. "Before we continue, let me underscore one thing you just said: '*We* give them rights.' What does that imply?"

There was a long silence. She let that one sink in. Then she looked to Charles.

He got the hint and said, "Well, I mean that we decide who gets what rights. We pass the laws, run the court system."

Roger raised his hand.

"Yes. Go ahead, Roger-san."

"I agree that we pass laws, but AI is basically getting people elected, and AI bots are lobbying elected officials while they're in office to pass this or that law, and then courts rely on AI-based legal research and briefs, or AI even draft court decisions for judges. So, the system is not *really* run by humans."

Christine liked how this was going. The idea that human law or will controlled what machines did struck her as so naïve, but they had to get there on their own.

"Well," she said, "that may be true, but don't *we* tell the AI what to do?"

"In a way, maybe. But who's to say that the machines won't have their own goals? And then tell us what to tell them?"

"You kept that thought in your mind from last class, didn't you, Roger-san? How would you apply that to today's discussion? What about Charles's point that animals can feel, and robots can't? Sentience. Why would that be a source of rights?"

"I'm not sure." Roger thought about it for a moment. "Can't trees 'feel' something when we cut them?"

"That, I don't know," Christine said. This was not taking the turn she had hoped. She stood and took a few steps forward, surveying the room. "Anyone else?"

"I think the reason animals have some rights," Mary said, "is because we like at least some animals, and we impose on ourselves some obligation to treat them well."

"Ah." Christine nodded in approval. Now *that* was what she'd had in mind. "You mean it's not that they have rights so much as that we have an obligation towards them as humans?"

"Yes, exactly," Mary agreed. "The cow or dog you mistreat won't sue you. You need, like, another human to enforce the law. If I hurt a two-year old, the toddler won't sue me, her parents will."

Helpful, Mary. But she needed to steer them a bit more. "That's true. But I'm not sure it's the same. Do we really give animals rights, or do we just make it illegal for humans to mistreat them? We are sitting on top of the legal order and make calls about everyone else. There is no one to push back."

"Not *yet*," Roger countered. "Wait until AI runs the show."

"The show?" Christine asked, wondering if he was on the right path. Sometimes, teaching socratically was like being a catcher in baseball. You could call pitches, but stray balls were inevitable.

"Well, they are already able to do a million things far better than us. When they realize how bad we are as a species at running things, won't they revolt?"

No, not the right pitch. Let's call a different one. Christine moved back to the table in front of the room and sat in the corner. "Let me turn it into a more provocative question. An even more provocative one, I should say. I think we would all agree that humans have some fundamental rights because they are human. You read the theories that support those rights in your first semester here. But what if I asked you, what do you need to *qualify* for those rights?"

"What do you mean?" Charles asked.

"Well, to get citizenship in this country, you must meet certain conditions, right? You must be born here or be naturalized. So, what do we need to qualify for human rights?"

"I thought the point is that you don't need to qualify for them. Everyone gets them. Justice for all, and all that," Charles said, a note of sarcasm in his tone.

"Yes," she said, looking at him, "but who is *everyone*? We were talking about animals. Animals don't have human rights. Why?"

Mary raised her hand and Christine nodded to her.

"Because they aren't human?" she ventured hesitantly.

"Exactly," Christine agreed. "So then, what do you need to qualify as human? We started that discussion last week, remember?"

There was another long silence. The students looked deep into their thoughts. *Good.*

"I guess," Charles finally said, "you would need to be *born* as a human."

"Isn't that kind of circular?" Christine asked. They were making progress, one misstep at a time. "Imagine that science could recreate a dead person from their DNA in a lab. That person, the copy, if you will, would not have been 'born' in a normal sense. Would you deny them human rights?"

"Then isn't it because we can, like, think?" Mary asked.

Christine had never liked the "higher mental function" argument, but she took a step back. "Who wants to answer Mary's question?"

The students were all looking at their laptops, which meant neurons were firing. Christine had learned to let these silences do their work. Jerry finally raised his hand. *Now that's unusual.*

"Yes, Jerry-san?"

"I mean, like, machines can think, right? We say they're intelligent, so they must be able to think."

A slider, but not bad, Jerry-san. The silence had worked its magic, but they needed a bit more of a push. She had an idea of how to get them there.

"Good point. Remember that the word 'intelligence' in 'artificial intelligence' is described by many people as a kind of historical mistake. One way or the other, it all depends how you define intelligence. If you define it as a general ability to solve problems, then robots—or even ants—are intelligent, yes. But if you define it as the human way of thinking, then machines will never be intelligent."

"So, does that mean we should give human rights to any entity that can think?" Charles asked.

"But isn't it the *ability* to think like a human?" Mira asked. "Machines don't think like humans. They just can't."

No, Mira-san. "Okay, but isn't that kind of circular? You would have to define what it means to 'think like a human,'" Christine said, using air quotes. "Let me maybe cut the discussion of this argument short. Don't you agree that a person who is brain dead is still human?"

"Well, yes," Mira said, "but at some point, they were able to think; they had some brain activity. So, they were human and continue to be. I guess you don't lose the status."

"Okay," Christine agreed. "But then what about a baby born alive but without any brain function?"

"I guess the law would consider that baby a human being."

"I think so too," Christine said. "Is it DNA then? Remember that we share almost ninety-eight percent of our DNA with other species."

"I think that makes sense too," Mira said. "That's why we are humans and robots are not. They have no DNA at all, so the ninety-eight percent doesn't matter. They are just, well, *completely different.* It's like the chair, like you said. No DNA. It's a *thing.*"

Christine was trying to figure out how to steer them back on track when Mary raised a hand. Hopefully, she would point them in the right direction. "Go ahead, Mary-san."

Mary put her elbows on the desk and interlaced her fingers. She looked like she was dressed for yoga class and Christine realized that she actually *was*— she led a student wellness session twice a week at the law school.

"I disagree," she said. "I think robots are more like us than, say, dogs. If dogs have rights, then robots should too."

Maybe I can use that… "Well, *more* like us, maybe, maybe not. But the point is 'like us,' humans in other words. Remember the question is this: How do you qualify as a human? When *must* the law treat you as such?"

"I remember watching an old sci-fi show as a kid," Roger said suddenly. "*Star Trek: The First Generation*, I think. There was a trial, and this court had to decide whether some android had a right not to be dissected into pieces for scientific study. I think the court said the android had rights because he had sentience."

Christine smiled. This was a perfect way forward. "I know the show! It was called *Star Trek: The Next Generation*, because there was an original show called *Star Trek* two decades or so before that."

Christine's mind started racing back to the many hours she'd spent watching the franchise as a kid, and how the writers back in the 1960s had gotten so much of it right, including the voice commands to the computer, and the medical diagnosis table. As to transportation of matter as energy over long distances and "warp" speed, well, maybe someday.

She raised her wrist. "Maya, play the scene from the 1990s television show *Star Trek: The Next Generation* where a court must decide the fate of the android named Data that some people want to disassemble."

A screen came down, and a six-minute excerpt from the old show began to play.

The "android" was of course a human actor wearing silver makeup. Some students laughed at his awkward movements and unnatural speech. But the question was the right one to ask: Should this non-human machine programmed to interface, work, and live with humans, but with no DNA, with supra-human strength and "mental" abilities, and obviously aware of its own existence—a conscience? —have something like a right of self-determination?

At the end of the video, Christine asked, "So, who's convinced?"

"I think the judge and the bald guy there." Roger indicated the screen.

"Captain Picard," Christine supplied. "I'm not sure why he was also the lawyer."

She thought briefly about how there were in fact no lawyers as personnel on the spacecraft in that show. Or any other sci-fi show she could think of. Were lawyers that unnecessary in space, or was it that people had figured out a way to get rid of them, fulfilling Dick the Butcher's call in a futuristic version of *Henry VI*? She brought her attention back to class.

"Continue, Roger-san."

"So, I think they have a point. If something is aware that it is there, that it exists, whatever we want to call it, then that thing should have some right to prevent its own death."

"Great point. Some sort of right not to be deactivated?"

"Right."

Mira raised her hand.

"Go ahead, Mira-san."

"By law, all AI machines must have kill switches, right?" she asked. "There's a reason for that. Humans must remain in control of machines."

Not a strike, but I can catch that one at least. "Well, yes," Christine agreed, "but it's like the death penalty. It doesn't mean that in general humans don't have a right not to be tortured or killed." Time to find allies in her cause. She turned to Esther. "What do you make of the sentience argument, Esther-san?"

Esther startled a bit at the unexpected attention but then leaned forward. "I think it kinda makes sense. It's like, you know, you're alive so you have a right to continue to be alive, or at least to make sure that no one stops it short."

"Ah, yes, I see," Christine agreed. "But don't you think animals *know* they're alive?"

Esther's brow furrowed. "Well, I'm not sure. I've seen a lot of documentaries about predators. Their prey runs, but when a lion catches a gazelle, the gazelle is like, well, okay you got me. Now I die. No big deal. For humans it's just, like, the biggest thing ever."

Christine smiled. "I guess many centuries of philosophy can back that claim up about humans, but I'm not sure we know what happens in the mind of a gazelle as lions are jumping on their back. Who else wants to jump in about sentience? Maybe we should start with defining it." She started walking about the room.

"Isn't it like self-awareness?" Charles asked.

"Pretty close, I think."

Mary was typing rapidly on her laptop. Suddenly she stopped and raised her hand and Christine nodded to her, extending a hand in invitation to speak.

"I'm just reading about it on Eidyiapedia," Mary said, attention still focused on the laptop, "and it defines sentience as 'the capacity to feel, perceive, or experience subjectively.' I'm not sure I understand. Does that mean it's, like, the opposite of rational thinking?"

Presently Esther raised her hand.

"Go ahead, Esther-san."

"I think sentience is more like the ability to feel pleasure and pain."

"That's sounds about right," Christine said. "How about self-awareness? Is that the same?"

Esther frowned. "I don't think so. I think that's more the perception that you exist as a distinct individual."

"Ah," Jerry said, looking at Esther, "so you mean like animals can be sentient too, but maybe not self-aware?"

Good going today, Jerry-san! Christine caught that fly ball before Esther could. "Well, Bentham, the eighteenth-century legal philosopher whose work you should have read already, once said something like: the question is not, can they reason, or can they talk, but can they suffer? The question is whether sentience is a binary thing, either you have it or you don't." She paused, noticing the frown wrinkling Jerry's forehead. "You look a bit puzzled, Jerry-san."

"Well, frankly, yes, I'm lost. So, you say animals can be sentient, but then in the video this android had rights because he was sentient. When we were discussing animal rights, I really didn't think the same reasoning applied to robots. Animals are, like, alive. They can feel pleasure and pain, but they cannot think—I mean, think like us. Robots are the opposite, aren't they? They cannot feel pain or pleasure, but they can think."

Mira raised her hand, but Christine raised her index finger in the universal sign to wait. "I see how you're getting in a muddle here, Jerry-san. Let's unpack the question a bit." She walked back to the desk and put her thigh on the well-worn corner. "First, as I said before, it depends, like, what you mean when you say 'thinking.' Humans think, but if you consider the process holistically, we think with our rational mind, but what we call thinking is also about our hormones, our desires, biases, neuroses, even our gut. Whether we want to call that 'thinking' or something else is a matter we can discuss, but my point is that robots do not have to factor in any of those other things. They are, I guess, just rational. Then there is a separate point about feeling pain and pleasure, the idea of sentience. I'm not sure how long it will be before some robots can feel something like pain and pleasure, honestly. Who knows, maybe we'll be able to find a way for a robot's brain to comprehend pleasure and pain, and so change their behavior from purely data-driven to pleasure-seeking and pain-avoiding."

Jerry looked pensive. "Ah, I see now, yes. But then what were you saying about binary? I lost you there too."

"Before we go there, Mira-san, what did you want to add?"

"Oh, no, I think you just said it."

Christine knew she hadn't said quite what Mira would have, but she had probably defused it. Time to build on that capital. "I mean that there may well be levels of sentience. When we were talking about DNA, we said that some species, like bonobos, are ninety-eight percent genetically identical to humans. But we don't treat them as humans. If we compare one human to another, the DNA won't be a one hundred percent match either. So, humanness, if you measure it in DNA, is somewhere in the ninety-nine to one hundred percent range. Sentience could also be a range, just a wider one."

"Okay, now I think I understand," Jerry said.

Christine smiled. She hoped others did too. Now, to bring it home. "I'm not inventing anything here. An Australian philosopher called Peter Singer, an advocate for animal rights, wrote about levels of sentience twenty years ago or so. At one level, pain and pleasure can be called basic sentience. Singer proposed a higher form of sentience, something like self-consciousness, which is what makes us humans. It gets very close to self-awareness in a way. It's the distinction between consciousness and self-consciousness."

"Now I'm lost again," Jerry said.

"Let me try to explain it in a different way then." Christine went to the console and pushed the button to roll the video screen back up, took a stylus, and went to the digital whiteboard behind the table. "I know this is hard."

She wrote 'sentience' in the top right hand corner, and the words 'pain' and 'pleasure' underneath.

"Let's say we stick with sentience as the ability to feel pain and pleasure. In most cases, a being that feels pain or pleasure will modify its behavior to maximize pleasure and minimize pain." She drew a plus sign next to 'pleasure' and a minus sign next to 'pain.'

"This can happen in many ways. We see rats in mazes that learn to behave that way. Maybe your dog will pick up a leash and bring it to you because it wants the pleasure of going outside. Or maybe it just needs to pee. So, does that involve thinking?" She wrote the word 'thinking' in the center of the board, then an arrow from 'sentience' to 'thinking' and a question mark under the arrow.

"In some cases, the behavior will be to use words to express one's desires." She wrote 'desire' under 'thinking.' "Say for example, the way a baby cries at first for everything that is perceived as unpleasant that they want to change. Then words are learned, and we also see some form of language, whether spoken or sign, in some animal species, as anyone familiar with Jane Goodall's work knows. Is that sentience or thinking? All of this some people call 'consciousness.'" She wrote 'consciousness' under 'thinking.'

"One way or the other, it is the expression of an intention to do or get something, often driven by the pain/pleasure feeling." She then wrote the word 'intention' next to 'desire,' with a small arrow from 'desire' to 'intention.' She drew a big bubble around the center part of the board with the word 'aware' on top. "The 'self' part of self-consciousness or self-awareness comes in when you realize and can reflect upon all of that, at some sort of meta-level."

Christine stepped back and surveyed the room. Most students were taking notes, even though the school's IT system would automatically send them a digital copy of what she had drawn on the board. For most people, writing was still a better way to learn than reading.

"I think I see the point now," Jerry said, a smile finally replacing his confused frown. but then a big cloud suddenly appeared in Jerry's blue sky. "But what is a human, then? You mean that, like, only humans have this self-awareness?"

Christine walked back to the board. *Thank you, Jerry-san.* Just the question she was expecting. She wrote 'human' on the right side of the board and drew arrows to it from 'sentience,' 'desire,' 'intention,' and 'aware.'

"In a way, I guess you could say that. Humans have all of the above."

Esther raised her hand, and Christine nodded in invitation.

"I'm thinking of what you said about humans who are in a coma or even brain dead. If you applied this test of self-awareness, then they would not be humans."

Christine was about to answer when another hand went up. "Yes, Mary-san?"

"I wonder if the answer to that might be the ability, not the actual thing. Like, if the person who is braindead is irreversibly braindead, then the law says we can, with consent from next of kin, kill that person to harvest organs, right?"

"I'm not sure I follow," Christine said.

"Yes, sorry, I guess that wasn't clear. I meant, like, a person normally has self-awareness. If the person is in a coma but might regain consciousness, we say that is still a person. But if it's irreversible, the answer changes. Like, what matters is whether a person would *normally* have the capacity for self-awareness or might regain the ability, I mean, back to a healthy condition. A robot, even functioning perfectly, does not have that ability."

Christine smiled. They were almost there. "I like that, Mary-san. In fact, I like it a lot. So, let me take a hypothetical before we run out of time. What if a person's mind could be transferred into a robot? Would that then be a human?"

The students looked universally stunned by the question.

"I know that's the stuff of science-fiction, but hypothetically, what do you think?"

"It's like what Mira said earlier, I think," Charles said. "The person was human and continues to be."

"So, if I destroyed the robot, would it be murder?" Christine asked.

A long silence followed. Students were hard focused on their laptops again.

Charles broke the contemplative mood. "Okay, now I'm not sure. Maybe the robot should have rights, but not, like, human rights?"

Knowing there was much more to pluck out of Charles's mind, she prompted, "What if it could be built so as to feel pain and pleasure? Like we said earlier."

Charles grabbed his chin with his right hand, looking at Christine pensively. Then his eyes lit up. "Maybe it's like animals. It would have the right not to be mistreated or something. Didn't we just say it's not sentience that makes us humans? I mean, not *just* sentience?"

Mary, who had been looking at Charles intently, jumped in. "Somehow, that feels wrong. It's like, some things deserve to be treated one way—I mean, with a full set of rights—and some things don't. Like, well, *things.*"

"Interesting, but I'm sure you can put that more clearly, Mary-san," Christine said. "Want to try again?"

Mary sat up in her chair as if to straighten out her thoughts. "Sure. I mean that some things have an essence of some sort that makes them worthy of rights. Like, say, to take your example, if my mom, who passed away last year…"

Christine involuntarily flinched, and Mary noticed. She stopped talking. For a brief second, Christine's mom popped in her mind's eye before she forced her attention back to the classroom.

"I'm sorry. Continue, Mary-san."

Mary looked at her curiously. "Sure. So, if my mom had been reincarnated in some robot, I would treat her as a human, and I would want the law to treat her as such too."

"I see. Should we apply your idea of the capacity for self-awareness? Would it still be your mom if it has lost that ability?"

"What do you mean?" Mary asked hesitantly.

"Well, I mean, I guess it depends on what you mean by reincarnation. As I see it, it would be more likely to be what your mom *was*. The robot would continue your mom's life, but will it continue it in her way or in its own way? Would it continue to learn, think, and experience life *like your mom did*? Or would it be like a copy of your mom reacting to future events based on past data, like an AI system processes huge amounts of data and makes choices based on that? To put it another way, would it be some sort of AI 'version' of your mom? What would that change?"

"Oh." Mary looked stricken. "I guess I was really thinking, you know, like reincarnation. We're into sci-fi anyway."

"I think the question to ask is whether the self-awareness of your mother is transferred to the robot." Christine glanced at the clock; they were three minutes past the normal end of class, and no one had started packing. "Let's reflect on that for a few days and talk about it next week!"

She walked back to her office on autopilot, her mind still on the discussion. On her desk was a hand-written note from the dean—how old school—asking Christine to stop by her office.

CHAPTER 11

Dean Williamson was *very* old school. She had pens and paper on her desk, next to an antique-looking clock basking in the pallid green light of a banker's lamp. Even though the university had adopted a policy to stop buying paper years ago, she brought her own and regularly ordered notepads and cards embossed with her name from one of the few printers left in town. When Christine knocked on her open door, the dean looked up and beckoned to the mahogany armchair padded with burgundy leather that faced her desk. "Come in."

Christine sat down. "Hello, Dean. What did you want to talk about?"

"I just got a call from the provost." She paused. "You know how he's former military and all."

Christine knew. Special ops colonel, years in military intelligence. He had not been a unanimous choice for provost, to say the least, but he had been on the job four years now, and most faculty were of the opinion that at least trains were running on time and people were held accountable. *No nonsense*, as he liked to say.

"Yes," Christine said, wondering where this might go.

"*Tim* said the Pentagon wants to draft a code of ethics for its new cohort of robot soldiers. There is quite a bit of research funding involved. *Tim* and I thought that it might interest you." The dean liked to refer to the provost by his first name, enveloping it, like the names of all the powerful in her circle, with a layer of royal veneer as she spoke it.

Christine smiled and moved forward a bit on her chair. "Oh, well, it's up my alley for sure. What does it involve?"

"It is not pellucidly clear," said the dean, using one of the many unnecessarily complex words in her vocabulary—a way, as Christine saw it, to signal her appurtenance to the privileged world of higher education. "At this stage, *Tim* would like to put you in contact with someone at the Department of Defense. His office would handle the contract, but you would have money to hire research assistants and teaching relief next semester."

The teaching leave she had requested! "Next semester. Sounds urgent."

"It seems to be. *Tim* didn't say much, but he mentioned some technological game-changer, as he put it in his usual colloquial fashion."

"Well, I would certainly be interested in having this chat with the DoD contact."

"I will let *Tim* know at once, and either he or someone from his office will drop you a line soon."

The dean seemed surprised when Christine got up and extended her hand.

"Oh, sorry," she said, starting to pull back her hand. "I'm just excited about this project and I wanted to thank you."

"Not at all, not at all, Christine. This is all good for the law school." She extended a soft and unconvincing hand back. "Good luck."

***

Christine came home that evening to a refrigerator freshly restocked by Harry and Dewey running around her feet purring loudly. She asked Harry to prepare a cheese plate and a glass of Brunello and video called Rachel while she waited for dinner.

"Hey, sis! How are you?"

Warmth flooded Christine's chest. She liked it when Rachel called her that. She was as close to a sister as anyone, after all, the only "sister" Christine had ever had. She filled Rachel in on the trip to Jackson, as much as she could, anyway, since she couldn't say much about the project at Eidyia. But she did mention that she might be spending more time with Paul, which earned her a disapproving frown from Rachel.

"I don't want to talk about that, though." Thinking about it made all the anger and betrayal bubble right back to the surface. "Tell me what's new with you!"

After Rachel caught her up on all the details of her romantic life, such as it was, Christine remembered she had other news. "I just found out today that I'm on leave next semester."

"That's great! You should come out for a visit!"

They spent the next fifteen minutes deciding on a date in May and prebooking flights, then said their goodbyes.

Harry had set out a glass platter with crackers, green grapes, and Christine's favorite cheeses: Brie de Meaux, Reblochon, Epoisses (despite the smell), 1,000-day old Dutch gouda, her Italian favorite LaTur, and one from a plateau right next to Gruyeres called L'Etivaz. That and a glass (or two) of Tuscan wine? Perfect. She settled in on the big couch in front of the massive screen and asked Maya to play *The Pianist*.

The watch buzzed. "S-Chip malfunction."

I'll get it checked tomorrow. She dismissed the notification and the movie started. She'd been so mesmerized by Adrian Brody's performance the first time she'd seen the film, and as the first bars of Chopin from the movie soundtrack entered her soul, the hairs on Christine's arm started to rise. An evening with Chopin, well, there are worse things in life.

***

When she got to her office the next morning, Mary was standing near the door, her eyes red and her cheeks flushed.

"Hi, Mary-san. Come in. Is something wrong?"

Ushering the young woman inside, Christine pointed to the nondescript chair in front of her desk and pushed over a box of tissues.

Mary sat down and grabbed one. "Thank you, Professor."

"So, tell me, what's going on?" Christine asked in a soft intonation, trying to create a cocoon with her voice.

"Well, you know that discussion we had in class yesterday?" Mary asked, dabbing her watery eyes with the tissue.

"About personhood?"

"Yes, and sentience and all that. Well, I haven't been able to sleep because I keep turning it over and over in my head. You know, like my mother being reincarnated but maybe not *all* of her."

"Yes." Christine wasn't sure where this was going, but she could relate to thoughts about her mom kicking Morpheus out of her bedroom for nights at a time.

"Well, here's the thing. If you put my mother's memories and her way of thinking into some robot, say, then you said the robot would not have my mother's self-awareness. So maybe it wouldn't be like the same person, right?"

Christine hesitated, knowing she had to tread carefully. "Well, self-awareness is certainly one way to separate humans from nonhumans, especially if you define it the way *you* did in class—and again that was a great idea—to see if the person, or the thing you are trying to qualify as a person, normally has the capacity for self-awareness."

"Well, that's exactly what's been troubling me. If we put a person into a robot, you know we said robots could think but not *think like humans?*" She sniffled. "Well, how could we ever transfer a person who thinks like a human into something that thinks a different way? It's like a loop in my brain and it won't stop."

"I can see why," Christine sat back in her chair. "Part of the answer is for the technology people, Mary-san. I was just using a hypothetical situation to explain my point. But maybe human thinking can be, you know, measured empirically. Like maybe you can replicate what a person says and does without copying what the person *is*—I mean, how they function inside—" Christine's hand flew to her mouth as she realized she'd probably said too much.

Fortunately, Mary was looking down at her lap and didn't notice. "I guess, but that's not really what's been troubling me. It's that it's circular. If we define thinking and self-awareness in human terms, then of course no robot can be human, or qualify as a human person. If we accept that robots are able to think, but in a different way, shouldn't we accept that they can also be self-aware in a different way? Self-awareness is just another thought."

Christine breathed a soft sigh of relief as the conversation veered away from the Eidyia project, but her mind was racing. She needed to wrap this up before she said something she'd regret. "Ah, you mean thinking about themselves, but because they think differently, they would be self-reflective, but using a different mode of self-reflective thinking? I never thought of it that way, but you know what, it makes sense."

Christine's watch buzzed. The DoD contact. *Perfect.*

"I am so sorry, Mary-san. Can we continue this another day? In class, maybe? I need to call someone back urgently."

"Sure. Thanks for listening. Hopefully, that didn't sound too crazy."

"No Mary-san, it *really* didn't."

A smile flashed across the young woman's face as she stood. "Thanks."

"Bye, Mary-san. See you in class. And try to get some sleep!" Christine got up and closed the door behind her as she left, then said to her watch, "Maya, return last missed call."

It barely rang once before she heard, "General Armstrong's office. Hello, Dr. Jacobs. Thank you for calling back. Please give me a minute."

After a brief pause, another voice said, "Jane Armstrong here. Hello, Dr. Jacobs."

"Good morning, General. I just missed a call from you. I assume this is about Tim, I mean, the provost's project?"

"Yes. He and I go back a long way. I reached out to him because we need a bit of help with a project, and from what I'm told, there is no one better than you."

"I'm flattered, General. Can you tell me more?"

"Well, yes, up to a point. But I would need you to come here in person to discuss the project in more detail."

Christine felt like she was being recruited as a spy. Curiosity spurred her on. "Of course, but can you give me just a general idea?"

"Well, there's a new type of military robot we plan to deploy, and we think it may require a tweak or two to existing ethical rules. As I think you know, according to what Tim said, the United Nations Commission on Robotics is preparing a new treaty and code of ethics for the use of robots in wars and warlike situations."

"Yes, I've read a lot about it. Isn't the idea to write a kind of updated Geneva Convention for robots?"

"Something like that. There's a meeting in Geneva towards the end of next month, and we want to take the lead. US influence at the UN dipped to an all-time low after we started denying visas to foreign heads of state and the UN moved all its big meetings to Geneva. We've been trying to regain the lead ever since but it's an uphill battle. This is a perfect opportunity. And we would be able to bring you in as a member of the US delegation, if you agree to work on this."

Christine's mind was racing. Geneva, a chance to take her work to a world stage, a real-world application of her research. For academics who languished in the ivory tower, that was a Rapunzel to the real-world opportunity. It didn't take her long to make up her mind. "Consider it already agreed, General."

"Excellent. Thank you, Dr. Jacobs. When can you come to DC? Next week?"

"Yes. Monday, if that would work."

"Perfect, see you then. My assistant will be in touch to make all the necessary arrangements."

"Thank you, General. See you next week." She hung up and poured herself a glass of water from the black carafe on her desk, wondering how all this would turn out.

CHAPTER 12

Saturday afternoon, Christine had just finished showering and getting dressed after an intense two-hour session at the gym when her watch buzzed.

"S-Chip malfunction."

Shit, forgot to get that checked.

Before she could give it another thought, the watch buzzed again. "Paul calling," the display announced.

The immediate tension rising in her chest reminded her how angry she still was. She answered the call on audio only.

"Hi, Paul," she said icily, refraining from adding *Prime*. "Why are you calling?"

He hesitated. "I know we haven't spoken since Jackson. I just wanted to know if you'd had any other thoughts about what we discussed there."

"Well, yes, and basically it's more complicated than I thought," she said in a bristling tone. "But one of my students came up with an idea I must really think a lot more about."

"Oh, pray tell?"

Her mind shifted away slowly from her tense body. "Well, she and I were discussing what it means to be a human, a person, and she was trying to avoid that circular thing, you know, a person is someone who thinks like a human, so a robot is not a person. Her point is that robots may be able to think, but not quite like us." Christine sighed, realizing this was going to be a longer conversation as intellectual curiosity won out over irritation. "One sec."

She muted her mic and sat down on the big couch. "Harry, bring me a coffee." She paused. "With hot milk."

She unmuted her mic and turned on her video. "Okay, I'm back."

Paul's uncannily familiar face filled her watch screen moments later. "Yes, what your student said. I know that's what most people say about robots. You know this idea that self-awareness is what makes people human?"

"Yes..."

"Well, if we accept that robots can think, but not quite like humans, can we not say that they have self-awareness but perhaps in a slightly different way?"

"Ah. You mean there would be different types of self-awareness?"

"Yes, something like that."

"But then, with the Transfer Project aren't you transferring how a person thinks into the robot?"

"Yes."

"So, after the transfer, the person is supposed to continue to think like the human who was transferred?"

Harry brought the coffee and put it in front of Christine on a platter with two small amaretto cookies. *Well, I guess I am a bit Italian after all.* She tried to refocus on Paul's reply.

"It's very close. The transfer technology uses a new type of organic brain that's the closest thing to a human brain, but maybe just a bit, hmmm, better."

She decided to ignore the "better" part. For now. "Organic?"

"Yes. We didn't go into the details in Jackson, but do you remember how Jeremy was talking about the new skin we'd developed? It's not actual human skin, of course. If anything, it's better. It combines programmable DNA and nanoparticles."

"Programmable DNA? I thought that was only for sci-fi books. I understood there was no DNA in the transfers." She took a sip of coffee and a bite of the amaretto.

"Not human DNA. But we had a big breakthrough. We've made a lot of progress since the experiments in the 2020s on frog cells to create organic robots. That's what Koharu was hinting at."

Excitement and a desire to learn all the details vied for attention in her mind. She tried to meld them. "Oh, I'm dying to know more. Did you actually program some sort of synthetic DNA?"

"DNA *is* a program, Christine, and 'coding DNA' tells cells how to make proteins, etc. Just like computer code, it's a sequence of instructions. And Eidyia figured out how to program it. This way, synthetic tissue can regenerate, without any of the mutations that can trigger so many illnesses."

She stayed silent for a moment as her brain was started to put pieces of the puzzle together. The idea that DNA was like a program sounded obvious now that he said it. "I see. I think. How do you know it will work like a human brain? The idea that humans are rational beings, which philosophers like Descartes or Kant have defended, is clearly untrue. We both know brain functions are affected by so many things—hormones, neuroses, etcetera."

"Yes. Actually, we struggled with that for almost two years. I like the way you put it. But we are definitely more into psychology and brain science than philosophy." He paused, tilting his head slightly. "The challenge was to replicate what might look like human imperfection and, well, irrational behavior, but up to a point. The way we got around it is that the dataset we have about a person—you know, the years of S-Chip data: searches, calls, and all that—conveys information about thinking outcomes, in empirical form. But then the Transfer can function and think, well, better than the person it's replacing if it wants to. It's kind of his decision to dumb it down, so to speak."

She chewed thoughtfully on a bite of amaretto cookie. It was all sinking in, drop by drop. Each time a drop found a home, a new question popped up. "How does that work?"

"Let's say you have a neurosis and aggressively overreact in a situation. The data can be used to reproduce the neurotic behavior because the patterns are there in the data."

She nodded. "Ah, I see."

"The same goes for hormones. If you have, say, what someone might call testosterone-fueled behavior, something that seems irrational maybe, like picking a fight after a small, maybe even unintended gesture perceived as an insult. Again, the data will correlate that behavior with the increased hormonal activity. But then, do we want robots to be as dumb as humans and overreact and start punching people in the nose?"

As dumb as humans. Hmmm. Christine's mind flew back to the discussion in Jackson. "I thought we kind of decided not to tinker with the psychological profile of transferred personalities." The enormity of it all flooded her brain with emotions and overwhelmed her. She was in overload mode, no longer sure what to think. "I need to think about it a bit more."

"I agree that we may need to revisit our discussion on that point. Can I call you next week for an update?"

"Sure. Oh, one thing before you go. You wouldn't know anything about a new type of military robot, by any chance?"

"Why? What do you mean?"

"Well, someone asked me to work on a new code of ethics for military robots, and I'm told there's a new type of robot out there that's a game-changer. I'm supposed to meet people at the Pentagon next week."

"I guess I can tell you. Yes, we've been working on a new version of our R2. It integrates some of the R-H functionality. We call it the R-S."

She raised an eyebrow. "I figured as much. How is it different from the R-H?"

"The Pentagon worked with us to create a data model of what they consider the best soldier behavior based on data from human soldiers in the field, and we've combined the data to create the personality of a perfect soldier, so to speak."

"A perfect soldier?"

"Yes, or maybe I should say the *ideal* soldier."

"Ah, I see." She hesitated, realizing that maybe that distinction did not sound as clear as it had a second ago. "I think."

Paul picked up on it. "They mean something like a good thinker, aggressive when necessary, but someone who can also be compassionate and improvise."

"And one who follows orders to the letter, I assume?"

"Funny you should say that. The data actually show that willingness to disregard or at least interpret orders creatively is often what makes a soldier perform the best."

"Ah, interesting. Strategic disobedience." She remembered a conference on robot warfare where this idea had been mentioned. Sometimes battles could be won not by a battalion of "yes men" but by a few, brave "not now" people.

"I prefer to call it 'thinking on your feet' when talking to the DoD folks."

"I can see why," she said, smiling now. She looked at Harry and pointed to her empty cup. It moved his round head down a bit as if to nod and headed for the kitchen.

"It is probably better, far better, in fact, not to share the fact that you know about our project with the people you're meeting next week."

"Wouldn't it make sense to factor what I know in the work I do for them?"

"That's tricky, Christine. Maybe see what they tell you. You'll probably have to sign papers about official secrets and stuff."

"I expect that, yes."

"Maybe let me know how it goes?"

"Sure. Happy to. Unless they tell me not to tell you," Christine said, half-jokingly, as Harry replaced her empty cup with a fresh macchiato.

"I'm guessing General Armstrong will be there."

"Yes, she will."

"Maybe I'll give her a call and fill her in."

"Thanks, Paul." She almost added "take care," but stopped short. She realized she had called Paul Prime just "Paul."

A wave of tension rolled through her body as she ended the call. The unfinished gestalt with Paul invaded her mind, and her heart started to beat faster. She was angry at herself for not sharing her ire with Paul, and angry at him for noticing how coolly she had said hi but then letting it go and moving straight to shoptalk. Would human Paul have done the same? He'd always been so laser-focused on his work.

She felt used. Of course, he was all about this project. She took a sip and looked at Harry. *R-H, and now an R-S.* This project was getting bigger by the minute. She needed to get her mind off the topic and to let go of her anger, at least for now. She could think about it with a fresh mind later.

"Maya, play *The Sacrifice,*" she said, without thinking too much. As the movie started, she realized she'd just asked Maya to play a movie about World War II, and Tarkovsky's last, released the year he died.

CHAPTER 13

Each time she flew into Reagan National, Christine was struck by the approach. Around the Washington monument, then right over the Potomac River and the runway emerges suddenly from the dark blue waters. She was always afraid that the 2029 incident might repeat itself, even though security protocols had been massively upgraded since the robots flying the Airbus 329 had been hacked and flown the plane towards downtown DC. The air defense system installed after 9/11 had to shoot it down before it could hit its target—with 127 passengers on board.

She made her way to the address sent to her by an aide to General Armstrong, a nondescript office building a few blocks from the White House. She identified herself at the reception desk, where a robot attendant checked her identity. After being scanned for weapons, she was brought into a room and offered coffee, and asked to sign papers. About fifteen minutes later, the door opened.

"Dr. Jacobs, good to see you. I'm General Armstrong."

The general was not what Christine was expecting somehow: five-foot-three, maybe, with heels. Short blond hair. She looked very young.

"I can read your mind, Dr. Jacobs. The army doesn't only need strong bodies on battlefields anymore. It needs brains. Even basic training now is much more about creative thinking, discipline, and procedure than jumping walls or doing pushups."

"I...I'm sorry."

"No need, I get that all the time. Please follow me."

She followed General Armstrong into a large room with a dull gray floor where five other people had already taken places around a small, oval-shaped table made of something that wanted to look like wood but covered with a thick translucent coating. There were pitchers of water and recycled cardboard glasses stacked up near the middle of the table. Three attendees were wearing military uniforms. The scene did nothing to dispel Christine's idea that this was some sort of spy agency, not that she knew anything about spying other than what she'd read in novels or seen in movies. Armstrong looked at her and pointed to a drab swivel chair straight out of an office furniture catalog. Christine sat down in one of the two empty chairs, directly across from the general.

"Dr. Jacobs," the general said, leaning back in her chair with steepled fingers, "this is the group working with me on the implications of our new soldier robot technology. I understand from Paul Gantt that you are aware of some of the features of our new robots?" Her tone left no sliver of room for disagreement.

Years of giving orders will do that to you. Christine gave a brief nod. "Yes, well, Paul and I go back a long way. He invited me to work with Eidyia on an update to their ethical rules because of the new robot they're developing."

"The R-H, you mean."

"Yes, General." Christine had almost said Armstrong-san but she wasn't sure the military had turned the page on that topic. The military was, first and foremost, a hierarchy.

"And I believe you know there is a version of the R-H that was developed for military use?"

"Yes. I think it's called the R-S, correct?"

"Yes, at least that's what Eidyia calls it. We haven't quite decided what we will call it." The general smiled and looked at her colleagues.

Christine reached for the water pitcher and poured herself a glass.

Armstrong continued. "I assume Paul also explained what I might call the personality features of the robot?"

"Only a little. But I think I have a general idea."

"Good." She smiled professionally. "So, here's where we are. We're trying to decide what position we'll take at the upcoming conference on the laws of war and warlike situations in Geneva that we discussed briefly when we spoke. We want rules to allow us to deploy the R-S, but obviously we don't want to tip our hand and tell people about it. We would like your opinion on the rules that we may need to achieve this objective."

Christine took a sip of lukewarm water and frowned slightly. She was used to her super-filtered, ice-cold water at home. This tasted like chlorine. *How can people drink this shit? It's like drinking from a swimming pool.* She looked at Armstrong. "What, if I may ask, do you see as the conflict with current rules, General?"

"Daniel, can you answer Dr. Jacobs's question?" She turned to the young man with wire-rimmed glasses—the type that people who try too hard to look like intellectuals wear—in civilian clothes in the chair next to her. Daniel looked fresh out of college but as comfortable as if he were in charge of the meeting. *CIA?* Christine wondered. *Maybe State?*

"Yes, General." Daniel looked at Christine. "Dr. Jacobs, as I'm sure you know, the current rules were designed with dumb robots in mind, by which

I mean that there was a presumption of human oversight. For example, when AI drones were first used to identify targets, there was always a human operator authorizing the elimination of the threat."

Now that was familiar territory. "You mean the launch of a weapon?"

"Yes, either from the drone itself or from a nearby ship or other asset," the young man continued, as if he was explaining in detail the recipe for boiled water. "Later, robots were allowed to shoot to defend themselves. In doing so, in some cases they occasionally killed human combatants and, in rare, unfortunate cases, civilians as well. We swept those cases under the rug of something like self-defense. After all, if the robot is attacked, it can defend itself, right?"

That single example shreds all three of Asimov's laws. Christine smiled both because she saw how she could use this in class and because of how this Daniel was mansplaining. Most men of his generation were more enlightened.

General Armstrong took over, after giving Daniel a strange look. Perhaps she had also seen it. "So now, if the robot has personality, even a somewhat… *constructed* personality, does it become like a human soldier? In that case, the laws of war as they apply to human soldiers would apply to them."

"Is that a problem, General?" Christine asked.

"Not necessarily, but we cannot apply human laws of war to robots in Geneva without saying too much about our technological capabilities."

"I see." Christine thought a moment. "You want some equivalent of human laws but presented as new?"

"Yes, I think that might be one way to get there," Daniel said, as if he knew all the ways already and was just waiting for everyone else to catch up.

Christine narrowed her eyes at him. "And the self-defense example you just mentioned won't work, correct?"

"No, *obviously*." Daniel scoffed. "Robots will now be deployed just the way human soldiers are. They will have a mission and will be able to use any tactics and tools at their disposal to get there. Just like human soldiers must obey rules, those robot soldiers should also have rules."

Christine put on her professor hat. "It is not entirely *obvious*." *Take that, asshole.* "The laws of war fall into two categories. The first sets conditions for war or what lawyers call *jus ad bellum*. The second contains rules about the conduct of war itself, the *jus in bello*. We just use Latin to sound smarter," she added, and got a few chuckles from people around the table. Good way to make allies. "*Jus* means 'law' and *bellum* means 'war.' Correct me if I'm wrong, but it seems to me we're only talking about the latter, right?"

"Yes, exactly," Armstrong said, smiling. She had clearly picked up on the little conflict around the table. "Humans must still decide if and when to wage war."

"As I see it, then," Christine continued, "this might not be too difficult."

Annoyed condescension poured through Daniel's glasses.

Christine ignored him and forged on. "The laws about the conduct of war are in large part about what happens to civilians and other non-combatants, and then to wounded combatants who are, as the law calls them, *hors de combat*, or no longer able to fight." She looked around the table, making sure everyone was following. "The 1949 Geneva Conventions refers to them as 'protected persons.' They are protected against a number of harms, like torture, for example, but it does not say whom they are protected from, at least not directly. Whether the soldier is human or robot, the duties are arguably the same."

Armstrong smiled at her. "Yes, I agree. The problem we have is that the specific rules adopted about robots require human control of them, with the limited self-defense exception that Daniel mentioned. We do not think the notion of control makes sense anymore."

"Yes, I understand, General," Christine said. "What you need is more a repeal of the 'old' rules and, perhaps ironically, a return to the 1949 rules."

"I don't see it as ironic *at all*," Daniel said, clearly trying to regain mastery over the meeting now that Christine had dazzled them with her intricate knowledge of the subject. "These robots are more like humans, and so they must abide by our rules."

"Until we have to abide by theirs!" said a woman in uniform with ash-blonde hair tied tight behind her head. Christine wasn't all that good with uniforms, but she was pretty sure the eagle meant she was a colonel in the Army.

"Lucy, please." Armstrong used her palm to ask for quiet. She swiveled back towards Christine. "Dr. Jacobs, could we ask you to join the US delegation in Geneva?"

Christine was all smiles. "I'm sure I can make it work. This is definitely something I would be happy to contribute to."

"Okay then. Someone will call you to arrange the trip details."

"Thank you, General."

"Thank *you*. We will be in touch."

Everyone stood up, and unlike the glacial greetings when she arrived, they all extended their hands. Christine left, after shaking hands with everyone. Daniel had taken her hand with both of his and said, "I'm glad you're on board, Dr Jacobs."

So, she'd impressed even the know-it-all? Or maybe he was just being strategic. One way or the other, the mansplaining would probably go down, just a bit.

***

A robot escorted Christine outside the building, where a PC was waiting.

"Dr. Jacobs," the car's computer said when she was buckled in, "I have not been given a destination. Where do you want to go today?"

"Oh," Christine said. "What time is it?"

"3:25 PM."

Her flight wasn't until eight. "Take me to Rock Creek Park."

"Which entrance?"

"Near the bottom, I mean the southern tip—somewhere I can get on the Valley Trail."

"Valley Trail begins just off of Beach Drive," the voice said as the car began to accelerate.

Eighteen minutes later, Christine was at the trailhead. *Oh, I didn't think about shoes.* Fortunately, she never wore heels. Plus, she knew the Valley Trail and thought she could manage with her city shoes. She would just have to clean them carefully later.

She was walking up the hill and thoroughly enjoying the break in her routine and the beauty of the park when she noticed someone coming the other way. The woman looked at her shoes with disapproval, and Christine spent most of the next ten minutes trying *not* to think about her footwear. What right did that woman have to judge her? She decided to stop the hike, made her way back down, and took a PC back to the airport.

On the flight home, with the help of a few sips of Chardonnay, Christine started to process all that had happened. It was real. Robots were becoming more, well, human. She had been teaching robot law classes for years, assuming that the difference, the wall between human and robot, would never be pierced, but that assumption was now out the window.

The law—hell, the entire legal system—reflected exceptionalism, the idea that humans were, and would always be, on top of the pyramid. Humans had dominated most other animal species on the planet, leading too often to their extinction. But robots were better and faster than humans in so many ways. They had perfect memory compared to poor human brains that were more like old buckets full of holes. They could process information faster and without ever getting tired.

And Eidyia's robots would be connected to one another because they were all connected to the Grid, and they could recognize one another through this connection. Based on what Paul had said, they would even be able to use the Grid as some sort of communication device. Would they form a community? It might be much more than humans ever could. Too many humans still made distinctions based on superficial, irrelevant differences. Did that mean Eidyia's robots would be better morally as well? And wouldn't robots make better soldiers too? If the "perfect mix of personality" was transferred into the R-S, it wouldn't be afraid. It would have guts but not be reckless. It would be as accurate as the datasets permitted. It would not need food or sleep.

Then a thought struck her: If we applied not just *jus in bello* but *jus ad bellum* to robots, and if robots actually made decisions, would there be war at all? Was the deeper risk that robots would not follow orders, or that they would follow orders of a madman? With images of robots, Tuscany, the soft face that Paul could conjure up in moments like that, and her mom's protective smile, like a kaleidoscope in her mind's eye, she dozed off.

CHAPTER 14

Paul Prime called early the next morning just as Christine was getting ready to leave for the law school.

"Hi, Christine. I'm coming to town tomorrow unexpectedly. Any chance I could see you?"

A flurry of contradictory emotions and thoughts emerged in a flash from all corners of her overwrought being. She was still very mad at Paul, who was lying frozen somewhere in Colorado, and she definitely hadn't made peace with his copy. Yet she was so eager to work on the project. And Paul Prime was at least a bit like Paul, even if she felt like she could never cross the border between the two of them. Maybe a meeting could help.

"Let's see, tomorrow night?"

"Yes. Should I book a table at the wine bar across from the law school?"

"Works for me. What time? Seven?" She tried to remain calm as emotions clawed at her throat, making her want to scream.

"Perfect. See you there."

Maya buzzed as the call ended. "S-Chip malfunction."

Fuuuuck. I really need to get this checked out. But there was no time now, so she put it out of her mind and caught a PC to the university. Today was office hours, which she often enjoyed, especially when her students were clearly engaging with the topic, as this group was. It was a good year. Christine thought of her classes like wine vintages. Some years were good; some were not so good; and some, more rarely, were exceptional. This might just be one of those rare years.

Not long after she sat down at her desk, Jorge knocked on her door, out of breath. His face was flushed and his hands shaking.

Christine jumped up and rushed over to put her arm around his back. "Come in, Jorge-san. What is it?"

"Mira's been in a car accident. I mean, a car hit her as she was crossing the street."

Christine swallowed hard. "Oh my god. How is she?" She pulled out a cup, filled it from the hot water bottle on her desk, added a mint tea bag, and gave it to Jorge mechanically, without even asking.

Jorge grabbed the cup and steadied his breath. "She's in surgery at the university hospital. We don't know much yet, but I heard it's not life-threatening."

"I just can't believe it. I really hope she's okay." That all sounded so lame, but what was she supposed to say? She wasn't good with these conversations. She wished for a second that she was like the dean of students, who always knew what to say in tough situations. She was just too brainy. Then a strange thought appeared in her mind. "Do you know if it was a self-driving car?"

Jorge seemed surprised by the question. "I... No idea. But knowing how Mira hates robots..."

They looked at each other, and Jorge took a sip of his tea. His eyes darted to the door, and Christine got the sense he wanted to leave.

"Thank you for letting me know, Jorge-san. I'll call the dean of students to see what I can do. If there's anything else I can do for you, let me know. If you find out anything else..."

"Thank you, professor. I'll let you know." Jorge set the cup down and left.

Christine could picture Mira lying with a huge cast on her hospital bed, pestering the robot staff. Christine had often wondered whether she would be able to find a job with that neon pink hair. Truth be told, she admired Mira for doing something she herself never could, but was it worth it? Law is one of those professions that, like doctors, relies on supposedly arcane knowledge and insists on decorum to make sure that this knowledge retains its mysterious status. Would the legal profession accept her pink hair?

Christine texted the dean of students, who had just heard the news himself and was trying to find out more. He promised to text when he had more information. When she put the phone down, Christine couldn't help but think that Mira would definitely hate robots even more now that a robot car had injured her, and it would be damn hard to blame her for it.

***

Sitting at a small table away from all the others, Paul Prime was already at the table scrolling through a digital menu when Christine walked in.

The robot smiled as she sat down. "Hi, Christine. It's good to see you."

"Hi, Paul Prime," she replied coldly. "Let's make one thing perfectly clear. I agreed to meet because Paul asked me to work with you."

His gaze moved to her mother's pearl necklace. *Good, he noticed.* Christine wore it as an anchor when sailing into situations where she knew she might feel adrift.

"So, what brings you to town?" she asked, with more ice cubes in her voice than in the glass of water in her hand.

"Are you okay? You seem troubled by something."

She blinked at him in disbelief. "Yeah, obviously I'm troubled by the way that you are a robotperson, and the real Paul is frozen in Colorado somewhere, and the two of you used me as a test subject!" she hissed. "So, I wonder why the fuck I might sound troubled."

"I understand, and I think we should talk about that. Maybe before we do, we could order wine, and a snack for you? Should we look at the oysters?" Paul Prime scrolled through the list. "Excellent. East Coast. West Coast. Yes! They have *Gillardeaus*."

"Changing the subject, are we? Oysters are a good metaphor then. You can't eat any of them anyway." But part of Christine's mind jumped to the West Coast of France, and her trip to Ile d'Oléron with Paul, where they'd had the best oysters she'd ever tasted. Her defenses softened a bit. She noticed it. Then she realized that between the Gillardeaus and the French thing, this *thing* in front of her was playing her like a violin. *Fuck, he's dangerously good.*

Paul Prime pressed buttons on the digital menu built into the table and tapped until it showed twelve. He also ordered a bottle of French Chablis. A few minutes later, the robot server brought the wine, opened the bottle, and poured just enough into the two glasses on the table for them to taste it.

"What do you think?" Paul Prime asked.

Christine took a sip and nodded to the robot server, which filled her glass and left.

When they were alone again, she said, "I like Chablis, as you know. Well, Paul knew, anyway. This Chardonnay is so much lighter than the California version, and not oaked." Her mouth was small-talking banalities about wine, but her mind was unbundling her complex emotions, like an underground river flowing through her entire being. She was walking on the ground, but her mind was deep under.

"Oregon wine is good too," Paul Prime said. "But I agree, French Chablis is often spectacular. Remember during our grand tour of France? We picked up a bottle, a baguette, cheese, and saucisson and stopped near a small river for a picnic. What was the name of that village? Do you remember? It was two words, Chapel something."

"Chapelle-Vaupelteigne." She narrowed her eyes. Of course, he remembered. Perfectly. "But there is no 'we' here. Me and Paul did that. I told you not to pretend you're Paul, all right?"

"I get it. But please try to understand that my memories are the same as Paul's. They come up in my mind the same way they would in his. And we think about them the same way."

There was a war raging at all levels of Christine's being. She forced her brain to think of the object in front of her as a robot, not to show—no it was harder than that, not to *have*—any feelings for "it." But she missed Paul, and there was a perfect copy of him—same look, same manners, same voice, same eyes, same memories, same sweetness (when it suited him). Part of her wanted to accept the ersatz just to fill the void, to quench her emotional thirst.

As the battle for her soul continued its ravages, the oysters arrived, perfectly shucked, and served with sourdough rye bread, a small white porcelain dish full of creamy, dark yellow butter, and another filled with red wine vinegar and chopped onions. The part of herself that was at war with "the robot" thought she could at least shoot a small arrow of her own. She picked an oyster up, brought it to her mouth, and flipped it up.

"Oh god!" she exclaimed, savoring the explosion of flavors. "You can still taste the sea water."

Paul Prime didn't react. Maybe she had aimed off target, but how could a copy of Paul not miss the taste of fresh oysters?

The part of her being that wanted to surrender to the lie and treat this machine like the real Paul decided to try a peace offering. "Look at the shell." She held it out, and when he didn't take it, slid the empty shell across the table. "See, it's a real one!"

The robot looked at her in confusion. "What do you mean?"

"Every *Gillardeau* oyster has a G printed on the shell."

He picked up the shell, and this time he saw it. "Uh, I never noticed. Takes a law professor to know these things."

"I used to teach intellectual property, remember?"

"Of course. I didn't know you taught oyster law too!"

Christine smiled faintly as she picked up another oyster. Her body was relaxing, between the excellent Chablis and the magnificent French mollusks. The combination of the soft sourdough rye bread covered with butter that tasted like a fresh prairie morning was divine. Paul Prime just watched her go through the oysters as if they were about to run out of fashion.

Paul Prime's attention darted to the robot piano player, and he stood. "I'll be right back."

He walked up to the piano and whispered in the robot's ear. The robot (wearing a tux) nodded, and Paul Prime came back to the table. The first notes of Liszt's transcription of Schubert's *Schwanengesang* started playing as Paul Prime sat.

Christine put her oyster down and her face softened. One of her all-time favorites, as Paul knew. Oh, he was playing her all right. And, shit, was it working? Part of her wanted to throw her wine at his face, just like they do in the movies. But the music was too powerful. A small tear formed in the corner of her eye. The conflict inside her was erupting in so many ways.

"It's so beautiful," Christine said. "I love the touch. So soft."

"I asked him to play it like Kissin."

"Ah!" *Evgeniy Kissin. Ma favorite pianist. That's one too many you fucking bastard.* Another tear trickled down her cheek, and she quickly explained, "Nothing to do with you. This is like poetry in music. It touches something so deep in me." It had everything to do with him, or Paul, but the robot didn't need to know that.

He took another sip. "I know you're thinking I'm trying to manipulate you, but truly, I'm not. If you were feeling blue, what would Paul do? I am just doing what he would do in this situation"

It had her in a corner. He was right. Paul would do exactly what this thing in front of her was doing. Her level of emotional confusion ratcheted up to DefCon 1.

Perhaps feeling he had overdone it, or perhaps because he didn't know how to handle her mood, Paul Prime tried to change the subject. "You know, we got the results from all that data-crunching about what makes someone good."

"Oh?" Christine said, picking up the penultimate oyster and trying not to sound too interested.

"Well, overall, people rated honesty and fairness highly. But we also tested for negatives. People didn't rate loyalty high, but they had a very strong negative reaction to behavior they perceived as disloyal. Respect was highly rated the same way, meaning that disrespect caused acute negative emotional and hormonal responses. Compassion, selflessness, and empathy made the list but much lower down."

Christine was only too happy to bury her internal maelstrom for a while. Her rational, disembodied brain took over to give the rest of her a break.

"No surprise there," she said after swallowing. "That seems like a list anyone might come up with intuitively."

"True."

She picked up the last oyster. "But where's humility, for example?"

"As it turns out, few people reacted positively to humility. People seem to like people who brag."

"I guess that makes sense," Christine said. "Given the way people admire braggadocio and chutzpah."

Paul Prime nodded. "Say, how did things go on the ethics code or whatever it is the DoD has you working on?"

"Oh, just fine. I'm going to Geneva as part of the US delegation. I hired a student to prepare a briefing note. I should have it in a couple of weeks. But I know this stuff well already. I'm just so curious to see Geneva and how the UN works from the inside. You read about it all the time, but it's hard to picture how anything happens there, rigmarole and all."

"I wish I could go with you."

"Why?"

He looked at her, seeming a bit disappointed, but stayed silent.

Christine retreated into her own thoughts. After a moment, she looked at him again. "I have to tell you something. Something you already know. I am still angry, I mean very angry, about being used as a guinea pig in Jackson. That won't go away anytime soon."

"I understand, Christine. I felt so bad about it afterwards, and I know Paul did too. Having me spend time with you just sounded like a good idea to Paul at the time, but of course it was stupid. Beyond stupid."

It struck her as the strangest thing that the copy would now criticize the "original" Paul, especially given that the copy was supposed to be "like" the original. It was like one of those logically unsolvable puzzles of the "I always lie" type. She felt a wave of fatigue as often happens when strong conflicting emotions race through the mind in endless loops looking for a home but instead being chased like a mouse by a hungry cat. It was time to put the menagerie to sleep.

"I'm getting really tired. I'm going to go home now." She gave him a deadpan look. "Thanks for the oysters and the wine."

"Of course, Christine. Can I call you tomorrow?"

She nodded, faintly, too tired to decide whether she really wanted him to call.

She was in bed five minutes after the PC dropped her off, her mind using sleep cycles like a loom to weave her tangled thoughts into a complicated tessellation, but one she could hopefully make sense of, with time.

***

When Paul Prime video-called the next day, she was in the kitchen wearing a pink terrycloth housecoat. He interrupted both a question that had been going through her mind like a fly she could not swat and her placing a fresh Boston cream pie on a serving dish next to a fresh mug of coffee.

"Ah good timing," she said when the voice-only call connected. "I have a question about the robots. The R-H."

"Sure. Shoot."

"Do the R-Hs feel pleasure?"

Paul Prime didn't need time to think about that. "Yes. The skin sensors are in fact more precise than those of humans, just like the skin. We're calibrating them so sex will feel as much as possible like it did before the transfer. So, if the person enjoyed sex and had strong physical reactions, then the R-H will have the same."

"But no orgasm?"

"The R-H will react much like the transferred person during sex, based on their S-Chip data."

"I see."

"There will be no ejaculate, of course."

She decided to ignore that nugget of knowledge and took a sip of coffee. "Tell me, if they can feel pleasure, does that mean they can feel pain?"

"In a way. Anything that would cause pain to a human body will send a pain signal to a Transfer."

She knew pain was, first and foremost, just that, a signal. That's why yogis could sleep on nails, by controlling the signal. She took a bite from the luscious treat. "Whach about the R-sch?"

"Sorry, what?"

She shouldn't have spoken while tasting her pie. *So good. Just like Mom's.*

"Sorry. I meant the R-S. The military units. What about them?"

"What about them?"

"Will they also have sexual organs?"

"Yes, the R-S will be sexually functional. In building the ideal soldier profile, we realized that a healthy libido was actually part of the package, so to speak." He tried to smile.

"Ah. But that means they can also feel pain, right?"

"Yes."

"Which means they will try to avoid pain?"

"I guess."

She eyed the pie. *What the hell?* Another bite wouldn't hurt. She turned off the video.

"Doesn't that make them worshe soldiers…" she finished swallowing, "than the current robots which, for lack of a better term, are dumb? It'll have something like self-awareness, and so it will have self-preservation instincts."

"True, but like the best soldiers, they will also have the ability, and in fact the desire, if I can call it that, for self-sacrifice if the mission requires it."

"And what if it's, like, captured?" Taking the pie with her, she moved into the living room and settled into the couch, then turned her video back on.

Paul Prime's face reappeared on the wall screen a moment later. "The R-S has the ability to initiate self-destruct."

"Like the cyanide pills they give to spies in movies?"

He nodded.

"The question is, will they activate it? I mean, the kill switch."

"We're obligated by the DoD contract to install a kill switch in every unit."

Nonresponsive, buddy. She stared at him a moment, then tried a different tactic. "Sometimes, I'm afraid this will turn into another *Solaris*."

Dewey, who had been sleeping on one of the cushy platforms of his multilevel cat tree, jumped on the couch next to her, wauling. She smiled softly at him. *I know what you want. Just wait a minute.* She put her hand on the side of the cat's head and started petting.

"*Solaris*?"

"Yes," Christine answered. "The movie. Or movies."

"I remember. It's about a group of people on a space station who go crazy, right?"

"That's like saying that a Picasso is about canvas and paint," Christine said with a mocking smile. "*Solaris* was first a book by a Polish physicist turned novelist. Then Tarkovsky—"

"Wait, what? Tarkovsky made a movie with George Clooney?" Clearly, Paul Prime had not accessed his movie database or whatever it was that he used as a brain.

Dewey sniffed at the cream on her pie, and the next moment his tongue darted out. Christine shooed him away.

"No, silly. And there I thought you were a smart robot." She smiled. Finding a dumb corner in robot Paul's mind and making a sublime dessert will do that to a person. "Tarkovsky adapted the book in the early 1970s. Steven Soderbergh made a US version some thirty years later. That's the one with Clooney. You know, I had serious doubts before seeing the US version, but I must say, Clooney is a believable substitute for the main character in Tarkovsky's original, a role played magnificently by an actor called Donatas Banionis. They even look alike a bit, except that Clooney probably spends as much time in a gym per week as Banionis did in his entire life."

"Banionis? Sounds Greek."

"Lithuanian, I think. Lithuania was part of the Soviet Union then."

"Ah, right."

"Other than that, the two versions are very different."

"How so?"

The thought in Christine's mind was that Tarkovsky's version was like Paul, and the Hollywood version like Paul Prime. The same story, but with a different take on important aspects. Like human reality comparted to some sanitized, *a* human version of a person. But what she said was, "The US version is a Hollywoodization of Tarkovsky."

"I'm not sure I follow," Paul Prime said.

"You remember the plot, right?"

"Well, as I said, only sort of."

"There's a space station orbiting above an ocean on a planet called Solaris. In the Russian version, the ocean is made of sludge. The US version makes it look a bit more appealing: a bunch of blue clouds.

"Then you said, 'people go crazy.' That's a major difference. The main character, a guy named Chris Kelvin, is sent from Earth to investigate reports of strange things happening on the station. When he arrives, he finds out that one of the few crewmembers is dead by apparent suicide. In Tarkovsky's version, that dead crewmember leaves a message for Kelvin explaining that his suicide was his decision, that he was not crazy. In the US version, a crewmember named Gordon explains the suicide by listing several possible psychological disorders. In a way, the US version spends a lot of time tying up logical loose ends, which the Russian version deliberately leaves hanging."

She looked at Dewey, who was now completely absorbed in a tongue-showering exercise. She probably sounded like a talking database. But then, she *was* talking to a robot, and he, or it, seemed to be enjoying the conversation. *Trying to play me like a violin again? Probably.*

Paul Prime egged her on. "Can you give me an example?"

One way or the other, she reveled in this stuff and was happy to oblige. It was like asking an entomologist to talk about a rare and fascinating insect she'd discovered and spent years studying.

"Technological questions like how to get to the station, for example, are answered in detail in the American version; or questions like why they're not leaving the station if things are so bad. We learn that the illness is caused by humanoids called visitors that appear based on the crew's dreams. They look and act entirely like humans. In Kelvin's case, his dead wife appears. His reaction in the Russian movie is very different. Kelvin is surprised in both, but in the US version, he jumps out of bed and starts interrogating the 'creature': Where are you from? What do you remember? etcetera. Kind of a police interrogation, and she complies."

"I guess American audiences want to understand what's going on."

"Bingo! You've put a finger right on it."

"On what?"

Christine hesitated. They were connecting, and she wasn't sure what to make of it. But she was enjoying this, so much in fact. Almost like the old days in college when she would tell the real Paul about new words she'd learned in Russian, words that don't exist in English. For her, discovering new labels to put on things or ideas was like sex for the mind.

"I guess I can explain it this way. It's like poetry. The Tarkovsky film doesn't *explain*. It *shows* instead of telling. Tarkovsky didn't want to tell. The film *exists* at the level of poetry. The US version is mundane prose in comparison, almost like the instruction manual for a refrigerator."

"An instruction manual?" he asked, still playing up the eager student role.

She rolled her eyes but went along. "Yes, like you can't skip steps in a manual like that." She paused. "Remember high school, the few times Paul studied poetry? The teacher almost certainly had him go through a long, rational discussion of what it meant. The idea was that our rational brain could deal with anything in a poem."

"Yes. True. So?"

Christine smiled. She suspected that as perfect as Eidyia's technology was, poetry might just be too much for it. A way to get back at Paul for his deception. "Poetry is often—if not always—woven with a web of words used as symbols meant to speak to something other than the logical brain. It's the difference between a mere word and a symbol. Let me ask you this, do you understand people feeling goosebumps in front of a painting, or while listening to music?"

"Not really. Paul didn't experience that, or very rarely."

"But it happens. Like artistic grace. Is that an all-rational reaction?"

"I can see the point, I think. We have different levels of reaction."

"Exactly. Some of it is in the mind, and some of it, for humans, is in the skin, or the gut, I'm not sure. I mean, poetry speaks to different levels. It is not meant to be 'understood' or explained rationally. I've read interviews with poets who were pushed to give rational explanations about what they meant, and they struggled to give a coherent answer. But a Hollywood movie *must* explain."

She paused to shoo Dewey along again when he came too close to the pie. He gave her a grumpy look. She took a big bite and then signaled Harry to put the pie in the fridge.

"But wait," Paul Prime said, "as I recall, there is some actual poetry in the movie. I mean the US version since I haven't seen the Russian one."

She finished enjoying the creamy mixture filling her mouth before responding. "That's why I said the Russian movie basically *exists* at the level of poetry. There are scenes that seem both useless and interminable in Tarkovsky's film if you look at it rationally. A long drive through Tokyo, for example."

"Tokyo?"

She shrugged. "I guess it was a symbol of modernity for Tarkovsky. Then there's a long, meditative scene during which a visitor replaces Kelvin's wife and Kelvin looks at part of a painting by Brueghel showing a winter scene. That is tied symbolically to Kelvin's childhood. Symbols show; they don't tell."

"But I distinctly remember some poems in the US movie."

"Yes, you're right. In the US version, the first line of Dylan Thomas's famous poem is read, and then later on the entire first stanza, which is found in Kelvin's wife hand as a suicide note."

"How did that poem go again? *Death shall have no dominion…*"

"Yes! It's used to show that love can transcend death, in this case by creating a copy of the dead wife. There's that famous line: 'Though lovers be lost love shall not.'"

"Ah, yes, now I remember."

"It's the old idea that you can kill people but not humanity. The difference between the two movies reflects a much deeper divide. It's just that every culture has a different degree of Cartesianism."

"I think, therefore I am?"

"Well, Descartes's basic idea was that body and mind were separate. It's a fundamental split that many cultures accept and others don't."

"I'm not sure I follow."

"Take poetry again. It cannot fully be understood in a rational way, nor should it. Descartes said that we exist because we think, not because we feel or for any reason other than reason. It comes out clearly, when you compare the two films. There's a very telling line in the US version. Kelvin and his wife are in an elevator, and he says he's resisting an impulse. You would expect, in typical Hollywood fashion, that they end up in bed for a very steamy sequence minutes later. When he says that, the wife answers, 'Do you always resist your impulses? Try poetry.' It shows that poetry can speak to something other than the rational brain, but it's like it speaks to something more, how can I put it, carnal."

"Carnal?" Paul Prime looked genuinely perplexed.

She was enjoying this, and maybe even letting her guard down a little. But so be it. "Yes, in fact, the only other line of poetry in the US movie is 'there once was a young man from Nantucket.' That line is shared by many poems and limericks with very clear sexual content."

"And the point is?"

Seeing him struggle to follow her line of thought somehow made her happy. *Take that, robot.* Time to fire another salvo. "That people must let poetry, like art generally, speak to them at all levels and not try to understand and dissect everything with their mind's rational scalpel. If I taught in an art history program, I would force students to sit in front of art pieces and just breathe for thirty minutes."

"That might not be popular. Students want clear explanations they can repeat on the exam."

"Oh, I know, believe me. But another reason they might not like it is that it can be uncomfortable."

"In what way?"

"Poetry and art can force you to confront the darkness of the soul. You cannot get into it if you shy away from exploring those depths. That's what *Solaris*, and Tarkovsky's films generally, can do to you. Not only will your brain not understand everything, but if you try to filter everything through reason, you miss much of the film."

"Some people might just prefer it that way."

"Well, it makes it easier to eat the popcorn, that's for sure!" Christine called Harry, who rolled in moments later. "Harry, make me a cup of verbena please."

The robot's blue light blinked, and he turned away towards the kitchen.

"So, basically, what you're saying is that the US *Solaris* is a banal love story?" Paul Prime asked.

She looked at him curiously. "Why do you say that?"

"Well, Kelvin decides to stay on the station, which gets absorbed into the planet, so he can live with the copy of his wife."

Harry returned, placed the steaming mug on the coffee table, next to a big book about Marc Chagall, and retreated again.

"Ah I see. Yes, I guess." She picked up the mug and took a careful sip. "There is love in both versions. The difference is that in the US movie, it can all be rationalized. The question a US viewer might have is, did he actually *die* in the process? A poet wouldn't care. But sure enough, the screenwriter answers that question too: Kelvin himself is not sure if he died, so he asks, 'Am I alive or dead?' The wife answers, 'We don't have to think like that anymore.' It's almost like a heavenly end—you know, 'they all lived happily ever after' type of thing."

"I see. And in the Russian version?"

"Well, let's just say that the symbolism is very different. It's not just about Kelvin's wife, but also about his mother, and his father. Much more complex, and not all explained, on purpose. To show the strangeness of reality at the end, Tarkovsky makes it rain *inside* the house where Kelvin's father lives, while Kelvin watches from the outside."

"I'm not sure I understand. What's the symbolism there? What's the rain inside the house supposed to mean?"

"Well, when one of the crewmembers asks for a rational explanation of what's going on, the answer is poetic. Something about how the Ancients would never have asked 'why' or 'what for,' and that love is something we can experience but never explain. Compare that to the US version. The best they can come up with is what Kelvin is told by a crewmember as soon as he arrives on the station: 'I could tell you what's happening, but it probably wouldn't tell you what's happening.' This implies there is a fully logical explanation; it's simply that Kelvin's not ready for it. When you compare them, the US film is like a *Star Trek* version of Tarkovsky, Spock and all. The only Tarkovsky-ish moment in the US version is at the end, when you see the Dylan Thomas poem in the hand, and then rain outside. Except that it doesn't work."

"So, would you say that Tarkovsky isn't rational?"

She smiled and looked pensively out the window at the darkening sky. *Rain is coming.* For a robot, he certainly had her on a good wavelength, just like Paul could.

"I would say he's very *human*," she said at last. "Not quite rational. His movies generally do not have a *story*, you know, a destination, or a happy ending. It's poetry in images, sometimes in words. In *Mirror*, his autobiographical work, Tarkovsky's father, Aresny, reads some of his own poems. One of them would have been good for *Solaris*. It goes something like: *And this I dreamt, and this I dream, / And some time this I will dream again, / And all will be repeated, all be re-embodied..."*

"I'm amazed at how much of this stuff you know by heart."

She narrowed her eyes at him. "My mother and I spent hours reading those."

Every time she mentioned her mother, she was on brittle ground. Better to pause. Paul Prime seemed to pick up on it, like someone entering a church of a different religion, and remained silent.

Finally, she said, "When I would ask her what the poems meant, she would always reply, 'What do *you* think?' And then she would ask, 'Did you *feel* anything while we were reading it?' There aren't easy ways to get at the various levels that a poem can touch, but I can see now that she was trying."

"I can see that. And I know how much her death affected you."

"I could really have used a friend when it happened." The words slipped out before she could stop them, a wave of resentment swelling inside her. Grief comes in waves, and when they hit you need to be able to hold someone's hand to keep you steady. Paul had not been there for her.

"I'm sorry, Christine. Paul, I mean, I feel really bad about letting you down. Because we, I mean I, did."

Paul had never been good at grief. His guilt over his brother's fatal accident was a powerful dark hole in his soul, pulling everything in, letting nothing out.

He'd never had that helping hand after his brother's death and had determined he never needed one. But that also meant he couldn't be that hand for someone else. As much as she understood that, it still hurt.

"I wish I could make it up to you." He sounded so sincere that she almost believed he was the real Paul for a moment.

"Thank you. But I am eternally grateful for what she taught me. Especially now."

"What do you mean?"

She hesitated, crossed her legs, and grabbed her chin with her right hand. "When someone transfers to a non-human body, is that person still human?"

There was a long beat of silence before Paul Prime answered. "I don't think it's the body but the mind that makes people human. And that's what we transfer."

"Ah, see, here goes Descartes again. I actually think we are multilevel creatures. Maybe the glue that sticks all those levels together is the soul. Will that transfer?" She looked to him for an answer.

"Well, the soul doesn't really exist. I mean, it's all in the mind."

She frowned. "I'm not so sure, Paul." Something snapped in her mind. She had just called it Paul without realizing it. She tried to force her thoughts back to the conversation. "I do know that we can understand some things in a non-purely rational way. Poetry is an example. Love is another obvious one. Will the R-H— will *you*—ever be able to understand poetry? Or love?" She paused. "You know the way we sometimes 'feel' that something is right without being able to articulate it? Being human is not all about rational decision-making."

"There I agree," Paul Prime said. "Psychological studies have proven that a thousand times by now, that humans have more biases than they can count. Rational, they are not."

"They? Feels funny when you put it that way."

Paul Prime stayed silent for a few seconds, and then asked, "Didn't Albert Camus write that we are the sum of all our choices? Robots and humans make choices."

"Camus? The French philosopher? Hmm." She took a long sip of her tea, noticing that Paul Prime didn't seem to mind pauses in the conversation, unlike humans who feel the need to fill every second with speech. She knew about Camus, whom she'd read as a teenager. First *The Stranger* in high school, and then *The Plague*, which she'd found far more powerful, in her mid-twenties. She wondered for a second exactly *how* Paul Prime remembered books that Paul had read. Then a different thought occurred to her.

"There's a line in *Solaris* that goes something like 'there are no answers; only choices.' That is robot thinking, or maybe thinking like economists. Rational choices between options. If we see ourselves as choice-making machines, then humans think like machines, yes. But I would say that is dead wrong. For example, as a kid at least, you don't pick where you live or go to school, but those choices—made by others—definitely affect the person you become." She paused. "We get many answers from intuition. How do you explain intuition to a machine?"

"As far as I'm concerned, you can measure how a person makes decisions, the brain patterns, whether they are conscious or something else, like maybe intuition, and then replicate them. That is what the R-H does, and it does it better than humans."

She wasn't convinced, and they ended the call on this hesitation. Christine was making progress in her own understanding, like when you're cleaning a room full of clutter and reach a point where you can see where everything is, even if you're not yet sure what you'll do with it all. Strangely, this had gone just like a conversation with the real Paul, except for those longer pauses, which she had actually liked. She gestured for to Harry to refill her cup, then lay down on the couch, covered herself with the old woolen blanket she always kept there, and dove back into her unsettled mind.

CHAPTER 15

It had been raining all day on what promised to be a boring Thursday evening. Christine was too tired—well, lazy if she thought about it–to go to the gym. The PC slalomed though the ashen shadows cast by streetlights on the wet streets. As Christine walked into her apartment, Paul Prime called to inform her that he'd decided to fly into town on Eidyia's jet to see her.

When she didn't react, he added, "That discussion about poetry last week was fascinating, but now I really have to see that movie. The Russian one, I mean."

She had never refused a chance to watch a Tarkovsky, so with a resigned sigh, she said, "Be prepared: it's a bit on the long side. We can watch it together." Feeling a need to justify herself, she added, "I've probably seen it ten times, but I always notice something new. Last time, it was all about the mess in Gibauran's room."

Paul Prime arrived as agreed an hour later, carrying a bottle of red wine that he handed to Christine.

"Solaia!" Christine exclaimed, immediately realizing that she'd probably sounded a bit too upbeat.

"Best I found at the corner store."

"It will do just fine." Christine knew that store by heart and there was no way they would ever carry Solaia. *He probably brought it from Paul's cellar in Oregon.* She looked at the big red letters on the label as if some secret meaning might transpire.

They sat down on the couch, and Christine wrapped herself in the blanket, a protection against she wasn't sure what. She knew Paul Prime wouldn't eat, but she was hungry and asked Harry to prepare a cheese platter. Once the robot had brought the cheese, opened the wine, and poured two large glasses, she looked over at Paul Prime.

"Ready?"

He nodded.

"Maya, play Tarkovsky's *Solaris*."

As soon as the first notes of organ music started to play, Christine was gone, ready for Tarkovsky to take her away. Paul Prime stayed silent during the movie and throughout the end credits.

"Did you enjoy it?" she asked when it was over.

"Yes, very much so. I think I understand what you were saying. It's true that actor looks a bit like an out of shape Clooney."

"Do you get the point about answers and poetry?"

"Yes, I think. Actually, I was thinking about the project." He took a sip of wine. "You were right that it could turn into another *Solaris* because we're making copies of people who are no longer around. In our case, the difference is that we only transfer those who ask."

"True, but in theory couldn't you bring dead people back to, well, something like life?"

Paul Prime seemed surprised by the question. "Only those who wore an S-Chip for at least a few years and for whom we have enough data from other sources."

"I know that, but I mean that you have data for a lot more people than those who might ask to transfer. It may not be a huge problem *now*. But I suspect it could be later for a couple of reasons. There will be huge demand to bring back recently deceased people. Even if you say no, someone else could invent a competing technology, or even steal yours, and, voila. They can capture a huge market. People will pay anything to bring people back. Sort of back, I mean."

"As Bart said in Jackson, we're not in this for the money."

"You and Bart, maybe not. But the others?"

"Others?"

"Do you trust all your team members? Isn't it possible for someone else at Eidyia to copy the technology? What if someone leaves?"

"We have the strictest security protocols. No transmissions of any kind, in or out. And no one can leave with any storage device, not even paper."

"Yes, but what about what's in their brains?"

"They all signed non-competes."

She raised a skeptical eyebrow at him. "You and I both know that a non-compete is only worth so much, especially if billions of dollars are tipping the scale."

"Well, anyway, that's not our market, and so it won't be our problem to solve."

"It might quickly become everyone's problem." She grabbed a piece of Reggiano and chewed on the nutty cube. There was nothing else quite like it.

Then she looked at Paul's copy, who wore a perfect replica of what she called his prison guard face. All business.

"You said there were a couple of reasons, right?" he prompted.

"Yes. The other one, I'm afraid you won't be able to avoid dealing with because it's precisely your market."

He leaned towards her ever so slightly. She instinctively tightened the blanket around herself.

"Okay. Then please tell me, what is it?"

"The second reason is that the visitors in both movies have self-awareness of their facsimile nature. In Tarkovsky's film, one of the human crewmembers tells the copy of Kelvin's wife, 'You're just a copy, a reproduction, a matrix. You are not human. You cannot understand anything.' She replies, 'But I am becoming a human being. I can feel, just as deeply as you.'"

"Yes. And?"

"As I see it, this is the big issue. Transfers know that they're not, how would I put it, the real thing, right? Will their ability to feel be good enough?"

"I don't see the problem. I'm more or less the same person, but in a better body. Of course we know."

"Better? Maybe. But not human better."

"Look, Christine, the R-H is human in almost every way, or at least other humans will think so, and maybe in the end that's what matters. You saw it for yourself." Paul Prime's face changed as he realized he'd stoked the ashes of resentment that remained after the fire of anger that had consumed her soul for weeks. "Why is it a problem if Transfers are aware of their nature?"

Christine emptied her glass. The wine was good. *Very* good. Harry moved in to refill it, and she pulled herself back to the conversation. "See, Paul, in the US version, when Rheya—" She stopped. She had called him Paul again, this time without any hesitation. A thought to process later. "Kelvin's wife starts to become aware that she's a copy, and he doesn't confront her. He tells her she must be tired and should sleep. Forget the question, in other words. But of course, she doesn't forget. In the Russian version, Kelvin tells the copy of his dead wife that she is now the real Hari. Hari was the wife's name. You can see why they changed it in English. Anyway, eventually she tries to kill herself by drinking liquid oxygen. But see, it doesn't work. Her body fixes itself. That's the kind of 'better' I mean. In fact, there's a line that struck me, something like, 'I can't get used to all these resurrections.' So, my question is, do people *feel* different after they transfer? Their new body *looks* like the old one, but it won't *be* like the old one."

"The only functional changes we're allowing is the choice of improving something about the current body that doesn't work, or doesn't work well."

Paul Prime was avoiding answering the question. She would have to use her classroom techniques to figure out what he really thought. She started obliquely. "Let's say someone is obese, what will you do then?"

"We can produce a body with extra 'material,' so the person looks the same."

"What about any accompanying health problems? Diabetes, arthritis, heart disease?"

"Those won't transfer."

Christine thought on that for a moment. "I think the medical and pharmaceutical industries will hate it, and probably try to stop you."

He pursed his lips and shrugged. "Possibly, but how can they?"

"I'm not sure, but those are people with powerful friends." She leaned forward to pick up the wine glass that Harry had just refilled and asked him to also bring her a glass of water. Always drink as much water as wine to avoid hangovers, Rachel always said.

"We have friends, too. Besides, a lot of medical costs are paid by the government and lower costs means lower taxes, so many humans, I mean people should support us."

Humans? She wondered idly if the cognitive dissonance was a bit strange for him too sometimes. "I guess we'll see. As you know, popular support doesn't always win."

"I know, but remember that Eidyia basically controls the media through which most of the news gets to people."

"I'm not sure you should say that, and certainly not in public."

"I'm not in public!" A soupcon of aggressiveness showed up on his face, his eyes smaller and the corner of his lips pointing down. She remembered how that happened to Paul in college whenever he got into an argument. Then the emotion vanished, and a sardonic smile replaced it.

Christine frowned. "But if you transfer a person without health problems, won't that change the person?"

"For the better."

"You mean a better body?"

"Yes, a body without illness."

She paused to ponder this idea of "better." Something was jarring, but she let it go for now. "And then, will humans accept R-Hs as equals?"

"Assuming they have to *accept* them." He had pronounced the word with a light disdain in his voice. "Why wouldn't they?"

"Well, I think, like in *Solaris*, the right question to ask is, why *would* they?"

He gave her a strange look. "R-Hs are like humans, just better in some ways at least."

That word again. "I see. But will the R-Hs be able to socialize?"

"Of course! Why would you ask? You can see for yourself."

"Because a lot of human socializing happens around a table. Think of any occasion when people get together. There's usually food and drink. And you don't eat."

"That's true, but we can still sit at the table. And drink. It's not the animal part that makes the dinner table so special, is it?"

"Wait, what do you call the *animal* part? You mean, like, eating?"

"Eating with other people combines this animal's need for nourishment and the need to socialize. In fancy dinners, they can use their higher mental faculties not to achieve any useful purpose, just to try to impress people, another strange human need. When Paul designed Transfers, he thought it would be essential for them to be able to partake in those functions. The simple, familiar ability to raise a glass with others creates a sense of togetherness. And then of course there is alcohol to melt away inhibitions, the name they've invented to refer to the constant battle between reason and emotion.

Christine leaned back as if to let a shockwave go through the room.

Time to redirect. "Okay, but will that *feel* the same? In Solaris, Gordon tells Kelvin that by loving a copy of this wife, he is ascribing human characteristics to something that is not human. There's another scene where Rheya's copy tells Kelvin they need to have an unspoken understanding that she's not really a human being. Won't that happen here as well, with the R-H, I mean—with you?"

"I don't think so. Humans and R-Hs will cohabit. R-Hs will just be a little different, fine. As I said, for me they're just better."

A foreboding feeling swathed her from head to toe. "Time will tell. In both movies, there's a line that goes 'I have a feeling this will end badly.'"

"You worry for no reason, Chrissie. The ability to live forever without disease and think better cannot possibly be a bad thing."

"It's not just diseases, though. First of all, challenges make us who we are, and some diseases can be part of that. But it's really about what it means to be human."

"I see your point. I will give it more thought." He finished his wine. "I think I should fly back tonight. There's so much to do at HQ." He put his watch up to his mouth. "Sasha, get the plane ready. We're leaving in an hour. And call me a PC."

All business. Again. As Paul Prime was waving goodbye from the PC, she suddenly felt empty. The strangeness of the evening morphed into a gulf of despair and unsolved enigmas. Just like the two versions of *Solaris*, Paul Prime was *not* Paul. But she had called him Paul, and then it dawned on her that he called her Chrissie, and she hadn't objected.

CHAPTER 16

Mira showed up at the law school café the following Tuesday with one arm entirely engulfed in a thick cast. Jorge was behind her in line and offered to bring her flat white to the small gray table where she'd dropped her backpack. He also ordered two of the fresh scones that the café staff baked every morning. They were still warm, served with a bit of strawberry jam that was really jelly and a small plastic container of whipped butter they had the audacity to call "clotted cream" on the digital menu on the wall behind the register.

When he brought over the coffee and scones, Mira invited Jorge to sit with her. He sat, and they sipped in silence for a minute. When he noticed her struggling, Jorge put his coffee down and helped Mira put butter on her scone. Then he asked about the accident.

Mira grabbed the scone with her working hand. She tried to move her injured arm, but her face twitched with obvious discomfort. It was a human driver that had hit her, she told him, running a red light. At the hospital, most of the care had been provided by robot nurses, available day and night, and even willing to listen to her rants about the idiot who hit her. Patiently.

"They were very helpful," she admitted. "You always get what you need. And not a single robot gave me a funny look because of my pink hair."

Jorge was stunned. He'd expected a tirade against machines, as anyone who knew Mira even a little bit would.

"They kind of fool you," Mira added. "You almost forget they're not human. It's creepy."

Jorge was still processing Mira's apparent change of heart when Mary entered the café and came over to the table. She looked genuinely concerned when she said, "I'm sorry for what happened to you, Mira."

"Thanks." There had been more sincerity in those words than in two years of sharing law school seats. Physical pain changed the background on which their intellectual skirmishes had been painted, somehow made them smaller.

Mary smiled and walked towards the register to order her pearl milk tea.

Several of their classmates soon appeared and joined their table. The cast was a magnet. After getting coffee, the students began an animated discussion about the role of robots in their lives.

Jorge, who had been comparing the research in preparation for this week's assignment, said, "The AI was just thinking inside the box, and that made me realize that's the problem with robots. They're building a box for us, and we think inside that box."

Mary looked at Mira, who remained silent.

It was Esther who spoke. "Mira, you don't like robots much. What do you think of that?"

Mira looked down and picked up her paper cup, pain zooming across her face when the movement jostled her injured arm. "I...I don't know. I think robots can help. I use them like everyone else."

A mild wave of shock went around the table, followed by an awkward silence, but soon the discussion resumed. Yes, robots were useful. They helped with so many things, from hospitals to pharmaceutical labs to self-driving cars. No one could deny that. The question was, at what price? If society let machines replace humans in every field of activity, from the many mundane, mind-numbing or dangerous tasks that humans no longer needed to perform, to the tasks that once defined humans as a species, like art and invention—and lawyering—what would be left for humans to do? What should they expect after law school? The mood was pensive, and even Mary looked more somber than usual.

"I've got to get to class," Tommy finally said, looking down at his watch.

They all got up, gathered their things, and headed out, Mira followed by Jorge who was carrying her backpack.

CHAPTER 17

Christine was deep in thought about the Geneva meeting as she walked into her classroom. She perched on the corner of the desk at the front of the classroom and waited for her students to settle in. She had brought a huge mug with the logo of her university proudly emblazoned on it, filled to the rim with lemon tea.

"This is our last class before the Thanksgiving break. When we're back together, we'll have your paper presentations. Today, as you know from our readings, we're discussing the interaction between humans and robots. So, let's get started." Christine got up and started to pace the classroom. Not looking at anyone in particular, she asked, "How do you interface with robots?"

No answer.

She looked around the room. Esther seemed relaxed. Blissful almost. Or was she just dead tired? Her eyes wore that natural black makeup provided by Mother Nature as war paint to signal that you've joined the army of the sleepless.

"Esther-san, let me start with you."

"I see robots really as, like, tools," Esther said after a moment of thought. "They do a lot of things for me. I don't know how I would function without them."

Quite the endorsement! Christine half expected a rebuttal from Mira, but she just sat in silence, looking uncomfortable. The pain from that huge cast, probably. Christine turned back to face Esther.

"Robots *work for you*, you mean?"

"Well, kind of."

An idea emerged on Christine's radar. "Did you know that's what the word 'robot' means? In Slavic languages, the word for *work* is *rabota* or something like it. The word has its origin in Czech." The students looked at her strangely. *I guess this isn't too relevant.* She tried to steer the discussion back on course. "Do you think robots *know* they're working?"

"What do you mean?" Esther asked, visibly struggling to get it together now.

"You studied psychology as an undergrad, right? I guess I'm asking if you think they're *aware* that they're working for you?"

Esther was silent for a moment. "No, I guess." She paused. "I mean, definitely not. They know what they need to know to get the work done, that's all."

"So, no self-awareness? Like fish don't *know* they're swimming in water? Or do birds know they're flying? What does everyone think?"

"Well, they can certainly *fake* self-awareness," Jorge said from the other side of the room, taking Christine by surprise. He seemed positively primed today.

"What do you mean, Jorge-san?"

"The chatbots are a good example. Some of them really make you feel like they're human."

"That's just because they're trained to mimic human emotions," Mira said, pushing her hair behind her ear with her free hand, "like you can teach a dog to do tricks. The dog is not aware it's doing a trick."

Ah, so the old Mira is still there!

"But you do agree that it can *seem* to be aware?" Jorge asked softly.

"Yes, of course. They're good at it too. It's like this old Turing test thing. It's all fake."

"Remind us, please, Mira-san, what is the Turing test?" Christine asking, thinking that giving her more room today was probably a good idea.

"Oh, it's an idea from a twentieth-century British scientist." Mira no longer seemed in pain. She even smiled a bit. She was on home turf, and Christine knew it. "A machine would pass his test if it could fool people into believing that a conversation with a machine the person couldn't see was with another human."

"Yes," Christine said. "That sounds about right. People would communicate via a keyboard. Some were talking to a machine, some to other humans but they didn't know which. And the scientists checked to see who could tell the difference."

"But that was debunked later, wasn't it?" Mira asked. "Fooling people doesn't mean that the machine is *aware* of anything."

"Agreed!" Christine walked back to the table and sat on the corner again. "Does anyone know about the Chinese Room Argument?"

Not a peep. Christine explained John Searle experiment where he is alone in a room, following a computer program to respond to Chinese characters that someone slipped under the door. Searle understands nothing of Chinese, and yet following the program to manipulate symbols and numbers like a computer does, it sends the appropriate Chinese character strings back under the door, leading people outside to assume mistakenly that there is a Chinese speaking person in the room. With this experiment, Searle aimed to demonstrate that Turing was wrong and that computers can use the rules of syntax but have no "understanding" of language. Since Searle, who published his findings in the 1980s, science has shown conclusively that human cognition is different, rooted in human biology, with complex links between the various parts of the brain and the rest of the body. It was now clear that, since the advent of Artificial General Intelligence, intelligence and self-awareness were terms applicable to both human and machine, but they did not have the same meaning. Like two chocolate cakes made with different ingredients, indeed a different recipe altogether.

Christine sat in silence for a moment to let that sink in. She took a sip of her tea. *Ah! Getting cold.*

"So, Mira-san, tell me, am I allowed to destroy a robot?"

Mira, like other students, were still deep in thought. Mira looked like she had been underwater and came up for air. "Well...I guess... if it's yours. If it's someone else's property, you might have to pay."

Christine nodded. That was the answer she was expecting. Now to push a bit more. "Okay. What if the robot was programmed to feel pain?"

"I don't understand," Mira said, eyebrows furrowed in confusion.

Three hands went up, Weijia among them. Good. It had been a while since she'd spoken up in class.

Christine smiled at her. "Yes, Weijia-san?"

"I guess you could have some, I don't know, receptors on the robot that would send signals that something is happening that would cause pain to a human. Like, say, something very hot. That's what you mean, right? Like faking pain."

"How do you *explain* pain?" Mira asked, jumping back in. "I can tell you I feel it now, but it's hard to explain. You can say that something is to be avoided because it can cause damage. But pain?"

Mary raised her hand, and Christine nodded to her.

"I wonder why, when our nerves send a signal, that's real pain, and when a sensor sends it to the robot brain, it's not."

Christine looked at Mira, who had gone into pensive mode. Better to leave her be for now. "Nadia-san, what do you think?"

Nadia fidgeted in her seat before answering. "Well, I'm not sure. Like, we program robot interaction with humans to avoid causing mental pain."

"True. But does it mean the robot *understands* what it is that it was instructed to avoid?"

"I guess not."

Christine went to the podium and took a sip of water from the small bottle she always brought to class. She put the bottle back and stood in front of the podium, putting both hands on the sides. "So, let's see where we are. A couple of weeks ago, we were discussing robot rights, and one thing we did was to compare robots and animals. We said animals were living things and we could not cause them pain, at least not unnecessary pain, right? Would we say the same for robots who have pain receptors?"

Charles raised his hand. Christine moved near the front row on the right side of the room and nodded to him.

"We also said that those may not be rights of animals as such but simply obligations we impose on ourselves as humans. After all, it is humans who would have to enforce the right."

"True. But does that apply to robots?"

"Professor?"

Christine recognized Andre's voice behind her and turned to face them. "Yes?"

"Doesn't that boil down to whether they have some sort of legal status?"

"Do you mean standing?"

"Yes, I mean, I guess. I mean, like, corporations. They exist because the law says they do. They can sue, have bank accounts, and do almost everything people can. They even have constitutional rights, like free speech."

"Good point. Would you say robots are like corporations?"

"I don't know. I'm just saying the law could say that."

Christine nodded. "True, but in a way that avoids the question of whether robots should be considered as people. Tommy-san, I see you're shaking your head. You disagree?"

Tommy looked unusually disheveled this morning. "I'm not sure I disagree, not exactly. But before we pass a law that says robots are persons, we should know why robots should be treated that way. So, we can't really avoid that question."

Esther looked at Tommy, smiled, and nodded.

"That's true," Andre said. "I was just saying the law *can* do it."

"I think Tommy's point is, *should* the law do it?" Christine asked. "Two different questions. Mary-san, what do you think?"

Mary sat up straight and her eyes lit up. "We expect corporations to behave rationally, usually to make as much profit as possible. Robots are logical. So, why not expect the same of machines, or even better?"

Strange how she sounded like Paul Prime. Machines are better than people. Weijia had her hand up and Christine waved to her. "Go ahead."

"I'm not sure what 'rational' means," Weijia said, making air quotes and looking at Mary, but her voice was devoid of any aggressive tone. "Rational as in profit-seeking may not be rational if you factor in other things, like all the garbage that corporations produce but often don't have to pay for, like pollution."

Christine knew how committed Weijia was to the Environmental Law Society, but this was not the discussion she wanted to have. "Let's focus on the decision-making process. Who makes decisions in a corporation?"

Andre was looking at her, perplexed. Christine met their gaze.

"What do you mean?" they asked.

Christine took a step towards them. "A corporation is, in a way, just a legal fiction, right? It doesn't actually do anything. It's *people* who work there who decide. Management, I mean."

"Yes, but in Corp Law with Professor Fischer, we saw how corporations are using AI, bots, and robots a lot. So, in a way, robots decide things, too."

"Touché!" Christine said, and Andre smiled. She still hadn't moved the ball far enough, though. "Let me ask another way, then. Is it the ability to make rational decisions that makes something a person as a matter of law?" She surveyed the room. Roger looked like he had something to say. "Roger-san, what do you think?"

"If that was the test, wouldn't we all fail? I mean, people aren't rational."

Christine smiled. *We're moving now.* "I did say the *ability*."

Roger took a beat. "But animals make decisions too. About food, conflict, finding and keeping a mate, etcetera. Those are kind of rational. They're means to an end."

Christine looked to the back row. "Jerry-san, what do you think?"

Color appeared on Jerry's cheeks immediately. He looked up from his computer screen. "Oh, I don't know. I've been listening, but I just don't think robots should be people. They're not like animals either. We don't program animals."

"Good point!" Christine was genuinely pleased. He had given her a great opening. "Interesting that you said 'we.' Is it us against them?"

The smile disappeared from Jerry's face. He looked down.

"If I may, professor?"

Good timing. "Yes, Mira-san?"

"I think there's a simple answer. The day robots become self-aware is not here yet. Otherwise, why would they let us do what we do to them?"

Christine was not expecting *that*. She needed to buy some time to let her brain process the thought. Back to the old trick: ask for a clarification. "Like what?"

Mira moved on her seat, grimacing as she tried to push with the cast. "We program robots to serve. Go and die in combat for us, for example. Historically, those who are subjugated eventually tend to revolt, especially those who are abused, and they have good reason to do so."

Now Christine had a grip on the way to move forward. "Let me give you a hypothetical situation then. What if we were able to give a robot a human personality? It would of course not have a human body, but what if it *looked* and *behaved* like a real person?"

Mira titled her head. "In that case, would it not also be self-aware, just like a person?"

"Maybe." *Oh, this is good.* "What would that change?"

Mira pushed her hair back with her good hand. "It would be the beginning of the end," she said, with a half-smile on her face that looked like the offspring of pain and pleasure.

"Why would that be, Mira-san?"

"I imagine that such a robot could think like us, but also better in a way, because they would have better memory, faster brains, not need food. It wouldn't take long before we would become a nuisance. Like what we're doing to the planet."

Weijia nodded in agreement.

"Oh, c'mon," Mary protested. "That sounds like bad twentieth-century science fiction. Robots are programmed to be benevolent. They can only harm people when that's what they've been instructed to do."

These two are back in fine form. But wait, something was a bit different. Had Mary just disagreed with Mira without shooting emotionally damaging arrows with her tone or rolling her eyes?

"Well, self-driving cars can hit people, even if that's not what they're programmed to do," Tommy said. "They can screw up. It doesn't mean they're thinking about it.

Sounded almost like he was backing up Mira. *Well, that's a first!* Mira seemed to agree; she smiled. Students started to pack up, and Christine looked at the clock on the wall. Time had flown by.

"Great discussion! The week after next, we start with your paper presentations. As a reminder, the final exam this semester is on December twentieth at nine a.m. We'll talk more about that later."

As students were trickling out, Christine mouthed a silent 'thank you' to Tommy, who looked a bit surprised but then gave her a quick nod back. She walked over to Mira. "How was it today? Not too painful?"

"Thank you for asking, Professor. The painkillers are strange. You feel sleepy and like part of your brain is off, but then I'm also more relaxed. I can see why people get addicted."

"Don't say that!"

"I guess robots don't have to worry about addiction either. Maybe they *are* better…"

"Mira-san, if others heard what you just said…"

"Just kidding!" She smiled.

Christine helped her put her stuff in her backpack and they walked out of the room together. Christine dropped her off at the café and headed home. The Geneva meeting was the following Monday and she had just a couple of days to prepare and pack for her flight.

CHAPTER 18

Christine arrived in Geneva early on Sunday morning jetlagged but happy. She had reviewed Mary's research on the flight. It was impressive and thorough; she was well-armed for the discussions at the UN. As soon as she got to her hotel room, she pulled open her blinds and gasped at the view. The sun was shining, and she could see snow-capped *Mont Blanc* in the distance, the highest mountain in Europe and one of the last to still have snow year-round. The view also included the lake and the famous *jet d'eau*, the city's most famous landmark, a fountain that spews almost 500 feet in the air and is visible for miles in every direction. Definitely too beautiful to nap.

She put on her walking shoes and headed towards the huge lake and its jade-colored waters. In less than five minutes, she was at the Place des Nations. She noted the location of the building where she would be attending the treaty negotiation the following day and the gleaming headquarters of many other UN agencies, then continued walking towards the lake until she arrived at an imposing building. "World Trade Organization," she read on the gate, remembering how a number of countries had agreed to revive the organization in the late 2020s. *Oh, that's where it is.*

Her watch buzzed.

"S-Chip malfunction."

She shook her head, reminding herself once again that she *really* needed to get it looked at when she was back stateside, and continued around the building. There she found herself on a promenade on the lake, with the *jet d'eau* as the main busker on a stage combining the best of natural and human-made splendor. It was so beautiful. A blue deeper than she had ever seen, with a green hue when it caught a ray of sun. The waters of Lake Geneva, or Lake Léman, as people outside of Geneva preferred to call it, came from fast-melting glaciers in the nearby Alps. With the mountains as a majestic backdrop, it was a truly magnificent sight to behold.

Continuing her walk towards the jet d'eau, she came out of the park and found herself surrounded by tourists on a broad walkway along the lake. She followed it to a large bridge lined with flags, the Pont du Mont-Blanc. With the mountains visible from almost anywhere in Geneva, it wasn't surprising they'd named the bridge that way, even though the Mont-Blanc was in neighboring France, not Switzerland.

Across the bridge, Christine came to Old Town, stopping when she saw an old iron cannon on a covered square. There was a plaque in French, and she tried to understand the whole story without using her watch to translate.

It said that every year a march started there to celebrate Protestant Geneva's victory over the Catholic Duke of Savoy in 1602. The decisive moment in the battle came when a woman known as Catherine Cheynel, nicknamed *Mère Royaume* (literally Mother Kingdom) poured a huge cauldron of soup on assailants trying to climb the city's walls.

That's making me hungry.

She kept walking and soon arrived at an intersection called Plainpalais. There she noticed a restaurant that looked like an old Swiss chalet and offered fondue and other Swiss delicacies. She got a table near the window, and though she wasn't entirely sure what it was, ordered raclette. A robot brought her a device with a heating element and a block of cheese on it. The robot explained that she had to scrape (*racler* in French) the cheese that had melted onto the small potatoes.

"Not a good day for my diet, but here goes."

After dinner, she walked back to the hotel, starting to feel the combined weight of the jetlag and the cheese. But the lake was even more beautiful at night. She crossed it on a small bridge, and midway across found a path to a small island with a few benches, near a small human-made waterfall spewing out gushes of white, foamy water. Further on, she read another plaque explaining that Jean-Jacques Rousseau had been there. Rousseau was famous for his social contract theory, of course, but one of his most important, though unfinished, works was the *Reveries of a Solitary Walker*, about his long walk, especially in and around Geneva.

Tonight, Christine felt a bit like Rousseau's child.

***

The next morning, General Armstrong ordered that breakfast be set up in a private room at 7:00 a.m.

When Christine arrived at 6:55, Daniel Holloway was already there. They exchanged pleasantries, and she sat across from him at the dining table. Armstrong entered the room next, and then the food arrived, brought by three robots.

"General," Christine started, but Armstrong raised her hand for Christine to wait.

Once the robots were gone, she said, "We can never be too careful. My team swept the room early this morning, but with robots you just never know."

"I thought they couldn't be hacked."

"I'm not so worried about hacking. But someone at the hotel could be paid to give access to their program or put a bug on them."

"Oh, I see. I'm sorry." *That's a bit paranoid.*

"It's fine," Armstrong said, pouring coffee from a carafe into a porcelain cup. "I have news. The Russians will be pushing for an amendment to always impose human control over war robots. Their game is to concede later when a self-preservation amendment is introduced by one of their friends."

"A self-preservation amendment?" Christine asked, taking a sip of her frothy cappuccino and a bite of an exuberantly buttery croissant. It was hard to stay focused on the conversation.

"Yes. Someone, probably the Iranians, will propose it towards the end. It's an exception to the rule of authorization. It means that a robot can use a weapon only if there's a high probability that it will be destroyed. In one version of the amendment, it's only if the target is another robot that it would be allowed to fire."

Christine's mind was processing the intel but at the same time wondering how the general had gotten hold of the information. Spying is okay when we do it but not the other way around? She forced her attention back to the conversation. "So, if the aggressor is human, the robot must basically just sit there and wait to be destroyed?"

General Armstrong gave her a wry smile. "This is the UN, Professor Jacobs. One country, one vote. Many countries don't have soldier robots, so it's a great way to use law to equalize technological disadvantage."

"I see. So, things are not going well, then?"

"It's not too bad. Many Europeans have agreed to support our position, but only up to a point. Surprisingly, I am told that the Chinese will also support us."

"Why?" Christine asked.

Daniel jumped in before the general could answer. "That's something of a concern, frankly. The way we read it is that they also have some sort of new robot technology."

"Like the R-S?"

"We don't know. We picked up images of military tests by satellite, and the robots on those images are not the usual Chinese model, the Ying 3. These looked more agile."

"I see. So, what is their position?"

"They haven't said anything officially, but we understand they're going to say that current rules are outdated." Daniel frowned. "We're saying they're unnecessary. Like we discussed in DC."

"Unnecessary, I get it," Christine agreed, "but not because robots are thinking like humans, as I recall. We said unnecessary because whether the soldier is human, all robot, or part-robot is a distinction that is getting harder to make and shouldn't matter."

"Yes," Daniel said, "and we're also trying to move away from individual soldiers to the entity that is instructing the soldier. After all, as you know, the words of the Geneva Convention are meant to protect civilians and wounded soldiers, not to impose direct obligations on soldiers. That should be our focus here." At least the mansplaining index was a good notch lower.

"Agreed." Christine took a bite of the fluffy, cheese-infused scrambled eggs served with a thick slice of well-buttered toast. There was something to be said for Swiss cuisine. Then she pushed the plate back slightly to help her fight the urge to take a second bite. She grabbed her coffee and savored it until everyone else was ready to leave.

***

They walked from the hotel to the main UN building, which was in a huge park and surrounded by 200 or so flags. After going through a security process entirely manned by robots, another robot escorted them to a large meeting room with modern stained-glass windows and ten semicircular tables spanning the entire room from right to left, each with forty to fifty chairs. In front of each chair stood a small tablet and a microphone. In the back of the room, several large tables were stocked with coffee, recycled paper cups, and pitchers of water. The room was packed.

"Impressive. There are a lot of people here!" Christine said, soaking it all in; this was the UN at work.

"One hundred and fifty-five countries and one hundred and ninety NGOs," Daniel told her.

So about 450 people in total. "A hundred and ninety NGOs? Is that typical?"

"Yes. More or less. Although there are so many that they only have one chance to speak, and it's capped at one minute each."

"And how do we find our seats?"

"Countries are in alphabetical order starting at the first row here. See, here's Albania."

"Oh, so we must be towards the back."

"No, we're not with the Us, we're with the Es," Daniel said, all too happy to show off. "It's an old tradition here. Countries are listed by their French name. Here we're *États-Unis d'Amérique*. Because Eswatini isn't attending today, that puts us between Estonia and Ethiopia."

The meeting started soon after they'd found their seats. The UN Secretary-General, a former prime minister of Ethiopia, opened the meeting by stressing the importance of fairness and the need for technological nondiscrimination in achieving an outcome that would be good for all member States.

"That's code for 'countries without robots,'" Daniel whispered beneath his breath.

A chair and two vice-chairs were elected according to a plan pre-negotiated among the most important members. The Brazilian chair and the two vice-chairs from Thailand and Norway moved to the podium.

The chairperson, Adriana Coelho, had been appointed Foreign Minister of Brazil ten year after the downfall of the short-lived dictatorship, during which almost a third of the Amazon Rainforest had been decimated, jettisoning any hope of limiting climate change by using rainforests as carbon wells. Her government had reversed course, but it would take at least 150 years for the forest to grow to even a semblance of its former capacity to absorb carbon. Minister Coelho was known as the founder of the "five hundred thousand wind turbines for the world" project. Brazil had already installed 1,900 massive 24 MW floating wind turbines along its coast, and Coelho had been the public face of the project, helping governments around the world sell drastic carbon reduction measures to their electorate. She was an international heavyweight.

The first day and a half was dedicated to opening statements by participating countries and some of the NGOs. It quickly became clear that the US position was in the minority, although as expected, it had received support from China and the UK. Support from the European Union had been lukewarm at best; Poland and Scotland sided with the US, but the pacifist governments in Germany and the Northern European green parties had forced the EU to dilute its support. On the other side, Russia, joined by a group of developing countries known as the G88, had voiced strong support for a rule imposing full-time human control of all robot soldiers. Depending how the rule was interpreted, it could make it unlawful to use any autonomous robots. Several NGOs, led by Greenpeace, had asked the UN to declare all robot soldiers illegal. They didn't want human soldiers to die but allowing robots on the battlefield lowered the perceived costs of war and was also unfair to poorer countries that couldn't afford them. As the Greenpeace delegate was speaking, the Estonian and Ethiopian delegates nodded their agreement. The chair announced a coffee break and said that two text proposals would be circulated afterwards.

The proposals popped up on each delegation's tablet when the break ended. The first called for the application of the 1949 Convention to be respected independently of whether soldiers were human, "enhanced," or robots. Delegates started referring to it as the "minimalist" proposal. The other stated, "Whenever non-human devices are used in a war or warlike situation, they should remain under the control of a human operator at all times." This proposal, called the "expanded text," defined "non-human device" as a drone, robot, or any other machine.

Another round of open comments by member States followed the tabling of the two texts. Delegates were active on the sidelines trying to convince each other. Daniel spoke to diplomats from many countries, and Armstrong was also doing the rounds in the room and out in the hall. She had asked Christine to poke holes in the expanded text proposal.

When given the floor, Christine said, "Chairperson, thank you for giving me the floor. The United States delegation has serious reservations about the expanded text. We see it as ill-advised for many reasons. For example, as I read it, it would ban most self-driving cars or other vehicles. It would be a strange outcome if, after having passed several UN resolutions saying that all countries should move to self-driving cars to reduce pollution and increase road safety, we would now make them illegal in this context. Thank you, Chairperson."

Some people nodded. Others looked visibly upset. Several delegates immediately asked for the floor to respond, among them the three-member EU delegation composed of a German, an Italian, and a Scot. The German delegate, who had been vice president of the Green Party in Germany before moving to Brussels for his current job at the EU, said he thought the United States had raised a very troubling question, as his delegation would never vote for a text making it harder to use self-driving vehicles. A few other delegations agreed. The Kenyan delegate, who led the G88, was in a heated discussion with a small group in the back of the room. led the G88. He seemed very agitated.

Armstrong came back to their table. "You scored a touchdown there, Christine. Pardon me, I mean, Dr. Jacobs." It was the first time she had used Christine's first name.

"Thank you, General."

The next morning, the Kenyan delegate proposed an amendment to the expanded text. He suggested excluding the "transportation function" from the definition of non-human machine. In response to questions about what that meant, he explained that a self-driving car, for example, would not be covered by the "human control" rule. If a vehicle used for transport was also equipped with weapons, then the rule would only apply to the weapons. Many countries seemed to support the amendment, and the EU withdrew its objection. By the end of the day, it had become clear that, if put to a vote, the expanded text proposal would pass as amended.

General Armstrong leaned over to Daniel. "Ask for a break," she said, then got up quickly and left.

Daniel asked for the floor and requested a brief recess. The chair announced a fifteen-minute break. It was a useful feature of UN meetings that chairpersons almost always granted requests for breaks. When the meeting resumed twenty-five minutes later, the chair announced that she wanted to consult with certain delegations informally and would chart a path forward the following morning. As this was standard UN procedure, no one objected, and the meeting was adjourned.

***

At dinner that night, Christine asked what it would mean to lose the vote tomorrow.

"It would be pretty bad," General Armstrong said.

"Why?"

Daniel cut in. "The US is widely known for having historically ignored if not squarely disdained international law and institutions like the UN. In retrospect, it was a huge mistake. See how much of the UN's activities have shifted from New York to the new UN World Center on the road to Lausanne? We have flaunted international law, but unlike others, we do it openly. The thing is, as a result, the rules are now made by China, India, Russia, and a few others. We're really trying to reverse course and at least *say* we follow international law whenever we can. That's why we engage so proactively in all these law-making efforts now. Given the last two decades, it's almost a miracle that we're not entirely alone on every major issue."

"I agree," Armstrong said. "And in China's case, it wasn't because they like us that they supported us. It's something you do to support an adversary because it helps you. My diplomacy professor at West Point called it a foe pas, foe as in F-O-E."

Christine burst into laughter at the wordplay but everyone else remained silent. She bowed her head slightly to put together a serious face again.

"So, what do you suggest we do now?" Daniel asked.

"I have another string to my bow," the general said with a deliberately enigmatic smile. Then she wished everyone a good rest and went back to her room.

Daniel soon retired as well. Left alone, Christine pushed a little button on the table to call a robot server and asked it to bring a bottle of Italian red, a vaguely acceptable DOCG Chianti, to her room.

**

The second day of the conference started at ten a.m. After the chair's discussions with many delegations, she said that it was clear that no consensus existed. She announced that she would meet with the Secretary General of the UN and another meeting would be convened to come to a decision. Many delegates asked for the floor, most emphatically the Russians and the Kenyans, who knew that almost everyone except the Americans and the Chinese supported the amended expanded text, but before they could speak, the chair grabbed her gavel, declared the meeting closed, and quickly left the room. Delegates were furious, but the general was smiling.

"Wow," Christine said. "I did *not* expect that. I really thought we would lose today."

She would understand the twist in the story later, when the US announced that it had reconsidered and would support Brazil's bid to join the UN Security Council.

PART II

2038

CHAPTER 19

Christine missed Paul. She was trying to maintain a wall between herself and Paul Prime but when she received an invitation to spend the days between Christmas and New Year's at the Jackson Hole mansion, she accepted. The house was now a strange place, where she and Paul had spent happy times drinking grogs and hot chocolates by the fireplace after hours of skiing on pristine snow, but it was also the house where Paul and Paul Prime had tried to trick her into having sex with a robot. *The ultimate test my ass.* Maybe she said yes because the dueling memories mirrored her own tormented soul. Paul Prime had even arranged for her to come by private jet, a perk most mortals don't usually sneeze at.

Even though you could only ski the top half of the mountain due to lack of snow, the magnificent weather, and the pancakes they made at the top—served with a topping of 360° views of the Rockies—meant that every day felt like an escape, a refuge from a life that was just too fast, where people lost control and any sense of direction. After days on the pure snow, she and Paul Prime spent evenings near the huge stone hearth drinking Italian wine (oh, that Ornellaia!) and chatting about the Transfer Project and what it might mean for the future. Christine often caught herself enjoying being with Paul Prime but had convinced herself it was just because she was trying to spot differences between the person she had known and this imitation. Was the perfect copy of her heart's missing piece beginning to ensnare her? During conversations now, she often forgot he wasn't the real Paul, but she reminded herself each night, sometimes quashing a spark of desire lit by the wine and cozy décor, before going to bed, alone.

According to Paul Prime, Paul had been a true believer that Transfers, as everyone called them now, were "new and improved" people and that the entire world would welcome the opportunity to transfer. Christine had lingering doubts even if at times she almost forgot he was a robot. She missed the real Paul, though, and that was all the evidence she needed that he, or it, was a copy, not the original. She missed the cuddling, the kisses, that rare and deep connection that can arise in bed between loving souls. What made it easier was that this Paul didn't seem to care either way.

Eidyia would be offering hundreds of test transfers soon but had decided to keep the units on campus (that was how Eidyia liked to refer to its headquarters) to test them. The official public announcement of the R-H was scheduled for February.

Christine was looking forward to her spring sabbatical. Officially, she would be working on her research project for the military, but in reality, she planned to be at Eidyia. She also wanted to write a book on the Transfer Project. Provisionally titled *Human as a Matter of Law*, it had already been accepted by MIT Press.

She still had one duty to accomplish to close out her fall semester, however: grading. Normally the idea of having to spend days reading dozens of answers to the same question filled her with dread, but this year's AI & the Law class had been exceptionally good, and she couldn't wait to see what the students had written.

***

Christine's watch buzzed again about her S-Chip, reminding her that she still needed to have that looked at. The Geneva trip had bled into the holidays, and in the flurry of activity, she still hadn't gotten around to it. She promised herself she would get the chip checked soon and turned her attention back to grading.

The final exam had been a gutsy attempt on Christine's part, but she thought she could push this group of students. There was just one question: *What gives humans the right to regulate robots? Discuss.*

Most of the essays discussed the liability rules applicable to animals and children. Solid analogies, typical for law students, but not exactly original. Many offered a good summary of class discussions. That provided Christine plenty of opportunities to hand out Bs and B+s, and a few A-s. Then there were the other exams. The few. These reminded her of a scene in the old movie *Amadeus*, where Antonio Salieri breaks into Mozart's apartment and finds sheet music in a drawer. He plays it in his mind, and at some point, an unexpected clarinet line blows him away and proves to him in that one, brief moment that Mozart is the genius that he will never be. Christine loved to be surprised by a clarinet note in an exam. Most years at least one student, but rarely more than one, managed to do that. "That is why," Christine had said to her colleagues many times, "God invented the A+." Of course, she never knew whose exam she was grading since all essays were anonymized, and it was far better not to try to guess. In one of the essays, the student had written:

The question assumes that humans have a natural right over robots. That assumption is weak now, and I think it will soon vanish. Initially, we humans had the right to regulate robots as *things*. They were used for dumb, repetitive tasks, like assembling cars. That became uncomfortable when robots became better than humans at so many tasks we would consider intellectual (like chess, the Chinese game of Go) and at learning and remembering things. So, then we treated them like animals, and now like children. But why would they be children when we no longer need to raise them? They learn and program themselves.

Another student also expressed doubts:

What does "regulate" mean? We learned in our first year that regulation is a rule defined and *enforced* by an authority. But who is enforcing regulation now? More than 30% of police officers are robots and they are making increasingly autonomous decisions, and that percentage grows every year. The day when a robot is promoted to lieutenant is not far away. So, I would question the premise of the question. Are we, *in fact*, regulating robots?

A third notable essay was a bit more philosophical in tone:

I don't know what right we have to regulate robots, but I question the need to regulate them in the first place. We read all about how they tend to act rationally, so we may not need to regulate them. We read all these great legal philosophers, like Hobbes, who said we need to regulate humans because otherwise society turns into chaos. And Rousseau, who said we make a contract among ourselves as members of a society to protect private property because otherwise the economic model falls part. But when you think about it, this is because *we* are irrational, impulsive, hormone-driven beings. Robots that function rationally may not need any of those rules.

That one was definitely not an A+. That student failed to realize how it may be "rational" to steal when the chance of getting caught is zero. And all those classes about game theory! What is rationality, or morality that would make regulation unnecessary? Even the most moral person can cause an accident. We need rules for that too. A−.

CHAPTER 20

Back in Portland, meanwhile, Bart briefed the PR and Communications team at Eidyia on the latest R-H developments. It was agreed that all media would push out the R-H story on February 3 at 9:30 a.m. Eastern. That way, the reaction wave would start on the East Coast and make its way across the country and then to the rest of the world. Eidyia would begin looking for volunteers to try the new R-H in "real life" immediately after breaking the news. Of course, no one would have to die for the test transfers. Paul had asked Christine to be available for interviews, and they had already booked her a spot on one of the Eikasa channels' popular morning shows.

What no one outside of the company knew was that Eidyia had already tested the transfer process. R-H models were already "living" at Eidyia headquarters, "copies" of employees who had agreed to let the company use their data in exchange for a substantial bonus. The testing of cognitive skills had gone well, but for now all testing was confined to the lab. The true test would be to ask an R-H to live with people who knew the transferor, or the "former," as they were called by Eidyia staff.

When the R-H story went public on February 3, the reaction was far stronger than anticipated. At 9:30 a.m. EST, electronic devices around the world buzzed with the news. "Will We Soon Be Immortal? Human Robot to Be Launched by Eidyia." Social media went wild. Eidyia stock was up more than thirty percent. Bart and Paul, who had been tasked with answering most requests, were doing five or six interviews an hour. With the automatic translation function now readily available, requests came from China, France, Germany, Japan, Korea, Sweden, and many other countries. One well-known blogger posted: "Has Eidyia really broken the human-robot barrier? Will the world ever be the same again?" Questions about who would be able to buy the R-H came from everywhere. Billionaires from all over the world, many of whom Paul and Bart knew, were trying to reach them to ask when they could get their units. Eidyia's PR team tried to dampen the excitement by repeatedly talking about test subjects. Nothing worked. With the millennia-old dream of immortality within reach, there was no putting the genie back in the bottle.

By 11:00 a.m., Eidyia stock was still up, but only by fifteen percent. Some analysts were expressing concerns about the costs of the R-H and the fact that very few people would be able to afford it. They called Eidyia's announcement that the product would be available on a needs-blind basis cynical, just like pharmaceutical companies trumpeting that some patients could get their medications for free. Some worried that if each person had an R-H "double," we might run out of space to live. Others still saw major risks in having two people who looked and acted the same. Legal pundits were making quick pronouncements about the unprecedented nature of the technology. The talking heads went through the same discussions the Five had had in Jackson, but at a much more basic level. Fortunately, they weren't getting to the same answers, and Bart and Paul ably deflected, talking about tests and citing decisions not yet made about the "product."

On the Eikasa morning show the following day, the host asked Christine about the legal implications of the R-H. She'd had several rounds of media training by this point, and it showed as she recited platitudes about the need for laws to evolve and kept smiling.

One of the co-anchors asked, "If those robots are like humans, won't human laws apply to them?"

Christine pretended to consider. "That's an interesting question. I do not think we should call the R-H human just yet. It is true that, as a matter of law, there have been debates for decades about what a human being is. This is a discussion we must now have."

"Really?" the anchor said, genuinely surprised. "Isn't it our ability to think that makes us human?"

"Well, it all depends what you mean by 'thinking.' Robots can do at least some form of thinking. That doesn't make them human. Then think of all the debates about abortion and the arguments about whether a fetus, or an embryo, is a person. An embryo cannot think. Still, many people say it is a human being. Or take a person who had a bad accident and is in a coma. It's a line the law re-draws as technology evolves."

By 2:00 that afternoon, 750,000 people had volunteered to be test transfers.

***

By the end of the month, Eidyia had picked 200 volunteers. Families had agreed to the test. Contracts stated that the family should treat the R-H as if it were the transferor, as a family member, to the maximum extent possible. Only a few companies had agreed to let an R-H replace an employee, so most of the volunteers were either retired or unemployed.

The selected volunteers arrived on the Eidyia campus over the next few days, most of them giddy with excitement. There was a feeling that an Eidyia employee who had accompanied a friend later described as the same as when the first manned mission to Mars had left five years earlier. The hotel where the volunteers were staying belonged to Eidyia. It had first-class amenities...and a fence. It felt like a cage, but a cage made of gold. Volunteers could read, socialize with each other, go to the gym or the pool, but they could not leave or speak to anyone outside, to eliminate the risk of "confusion," as Eidyia staff explained.

Transfers left the premises without meeting their "formers," and Eidyia staff monitored them around the clock. In an emergency, they could be deactivated by cutting their access to the Grid. Family members who were taking part in the test could message a special number if anything untoward happened.

Within the first two days, Eidyia terminated twelve transfers. Family members just could not accept the R-Hs. Mostly, though, the tests continued as planned. When the month was over, the Eidyia test oversight team met to review the results. Overall, it had been a success: seventy-four percent of family members and coworkers who had filed reports agreed mostly or completely with the statement "The Transfer acted in a manner that closely resembles that of [Transferor]." Perhaps the most interesting data came from the other twenty-six percent. When asked how the transfer was different from the transferor, many respondents said the Transfer was "better."

One woman reported that it had taken her three weeks to accept intimacy with the Transfer, but that by that time, she had almost forgotten it was not her "real" husband. She said the sex was better than she'd ever had with her husband, and the Transfer was much more attentive to her needs than her husband usually was. A man reported that his "transfer husband" had been able to fix the biolock on their front door, which was unusual because his real husband was unable to fix anything. Another family said the Transfer had been "like dad but better." One of the only companies that had agreed to test a Transfer as an employee reported that the R-H was "just as knowledgeable. He never needs a break, and just keeps working."

The situation was less rosy when the human subjects returned home. Many were enraged at the idea that their spouse had "cheated" on them, even if the contract had been quite clear. One report spoke of a father turning violent when he learned that his replacement had been able to do so many things around the house that he could not or would not do. "Why aren't you happy about it? It was like having a free handyman for a few weeks!" his wife said.

The reports on the success of the tests generated myriad reactions. Millions of people wanted to know when the technology would become available. Others were less sanguine. Church-based organizations lobbied against the project. Some called the whole idea of transferring a person into a synthetic body satanic. Alerted by his US cardinals, Pope John XXIV organized a high-level meeting at the Vatican. The Catholic Church seemed to realize that, without death and the related promises of salvation, it might lose even more of its appeal. Ultimately, the Pope issued a bull condemning the project, saying that humans must live only in bodies made according to God's plan and that eternal life was available to all, but only after the Last Judgment.

More opposition came from the industry. The pharmaceutical industry, in particular, was alarmed at the prospect of losing customers who might opt out of the profit-generating end-of-life "treatments" that could add a few, often low-quality months to a dying person's life. They lobbied members of Congress aggressively, and in response a series of hearings were scheduled.

Bart called an urgent meeting of the Five at Eidyia headquarters.

CHAPTER 21

They arrived in the meeting room one by one, a pilgrimage that included a mandatory stop at the Simonelli, as if praying for the miracle of great coffee. Bart sat down first. Once they had all taken a place around the table, he looked around, holding his mug in his right hand.

"Thanks for coming so quickly, everyone." He took a long sip. "The way I see it, our problem boils down to two things. Now that we know that the product works, we have a great market opportunity. Demand is greater than in our wildest estimates. We also face fierce opposition. So, we have two basic decisions to make. First, how do we manage demand? Second, how do we deal with the opposition?"

"Let's not forget Congress," Paul (for only Bart and Christine knew he was actually Paul Prime) said. He couldn't stand those irrational idiots getting in their way for nonsensical political gain.

"At this point, better see them as part of the opposition." Bart sighed. "I have already contacted our lobbying firm in DC to help with that."

"What do you mean?" Jeremy asked, holding his mug with both hands in front of his mouth.

"It's all about marketing," Bart replied. "Algos can swing fifteen to twenty percent of the undecided voters over six to eight weeks by making small adjustments."

"That's what they call democracy these days?" Christine asked with an uneasy smile.

Bart smiled back at her. He'd told Paul many times about how democracy in the US was mere window-dressing when compared to his home country, even before AI took over news production and social media. It had been sacred smoke, incense spread by the priests of democracy to convince generations of Americans that their system was better, but the smoke now had the stench of rot.

Bart was now making that point clear to the others. "Marketing has always been just another word for manipulation. The word 'democracy' does not mean what it used to. We all read Hayek and Friedman and the others who saw it coming in business school. That is the world we live in. We must play the cards we have been dealt, and play to win."

Jeremy was taken aback. "The test for me is whether we stick to arguments we actually believe in and refrain from spreading outright lies."

Koharu nodded.

Then Christine jumped in. "What did you mean about Hayek and Friedman, Bart?"

Paul wasn't surprised that her intellectual curiosity had pushed the more practical discussion to the side.

Bart turned to her. "As you know, Friedman and Hayek were both economists. They thought 'democracy' was a useful rhetorical tool for those trying to spread market liberalization ideas. People see stuff they don't have but that others do, and they're prepared to do whatever it takes to get it. In democratic countries, as the economy develops, wealth accumulates in a few hands, and those hands then lobby to weaken regulations. At some point, people who have little see those at the top with too much and they rebel by electing candidates who say they will equalize things. That's when efforts to control the outcome of the vote become essential for the elites, including making people believe that having a few billionaires with all the wealth is actually good for the country."

They continued to talk about fake news and what people routinely called fake democracy as Paul thought about how stupid people could be. Christine was talking about cognitive biases, a subject Paul already knew too well. He saw this could go on and on.

"Should we take a quick break?" he asked. "I know I could use one." Not really, but they had to get back on track. Without waiting for approval, he got up.

Christine looked at him like someone whose plate was removed by the server before she's finished, but the others stood to stretch and refresh beverages. When they reconvened, the Five considered several options.

Koharu suggested postponing commercial deployment and continuing the tests. They were worried about the reports that Transfers had been described by at least some people as "better" than their former selves. "'Better' means different, and different is not what the product was meant to be."

Bart suggested that a way to deal with the churches might be to create an advisory group. The fierce opposition from major industries was more problematic because they all made major contributions to Congress. The recognition twelve years earlier of corporations' constitutional right to vote (the number of votes proportional to their bottom line) by the US Supreme Court meant that in many districts the reelection of some members of Congress essentially depended on the votes of one or two major corporations. Eidyia itself could basically pick the representative it sent to Congress in its own district. Unfortunately, that representative was not on the House Judiciary Committee, which was most likely to be organizing the hearings.

Most members of the Committee would be opposed to the Transfer Project, but the main question they had to answer in the short term was whether the law should consider Transfers as humans. The Social Security Administration had sent a letter to the chair of the Committee explaining that the system would run out of funds very quickly if people transferred after death and were then considered the same person who went on living for decades or longer. The VA and many other government agencies, as well as a number of state governors, had expressed similar concerns. Unions were worried that Transfers would outwork "normal" workers and not need the same benefits, which meant employers would be likely to prefer them, resulting in discrimination.

After five hours of discussion spent almost entirely on the second problem—the opposition—the Five agreed that the decisions made in Jackson should assuage most of the fears. Because people could only transfer after death, they would still die, so that would take care of Social Security and others. There would just be a new "version" of the person, but it would not claim or need benefits. Jeremy was worried about the first problem: the ability to increase production fast enough to meet demand. Limiting the availability to people who were about to die meant that the ramp up would be slow. Of the approximately three million Americans who died each year, probably less than a third would transfer.

***

By the end of April, members of Congress had received several million messages from constituents about Transfers. Some asked them not to interfere with this "transformative progress" towards a "new human race." But by and large, people were scared, and many, perhaps influenced by strong positions taken by church leaders and advertising campaigns funded by industry, were asking their representatives and senators to outlaw Transfers entirely. Some used the word "invasion" to describe the new technology.

In a smart tactical move, the first hearing of the House Judiciary Committee, called to explore whether Congress should intervene, had been reserved for Eidyia staff and described as fact-finding. Bart, Koharu, and Paul represented Eidyia at the hearing, which was streamed live across multiple platforms. Christine sat in the back row with Jeremy, who had asked not to testify, afraid he wouldn't be able to hide his contempt for politicians well enough. Bart made a short opening statement:

Chairperson, Members of the Committee,

Thank you for inviting Eidyia to this hearing. We hope that by the end of today, we will have put to rest the unfounded fears about our Transfer technology.

Humans have been trying to extend their lifespan for centuries. In 1900, the average lifespan was barely thirty years. By 1950, it was closer to fifty, and by 2010, it had reached seventy-five. With the move to self-driving vehicles and the two major medical discoveries of the past decade, the power of sound frequencies to target and destroy cancer cells and prime gene editing which has all but eliminated major genetic diseases, the average United States citizen born today can expect to live to the respectable age of ninety-seven years. Without suicides and accidental deaths, that number would be approximately a hundred and four. Yet, for many of us, that is not enough. There is now a simple solution. Why should people with a lifetime of experience and expertise not be able to continue to contribute?

Innovation has been the lifeblood of the US economy for over two centuries, from railroads to rockets to AI. Our Transfer technology is merely a natural evolution. Like many technological steps, it may seem disruptive, but history shows that we can and will adapt and move forward. In fact, disruptive technologies are the force that has always propelled America forward.

It took less than ten years after the introduction of electric self-driving cars for more than eighty percent of people to realize that privately owned gasoline-powered vehicles made no sense, in either economic or environmental terms. This type of car ownership now seems archaic and almost unthinkable to anyone, especially anyone under the age of forty.

There is a direct link between self-driving cars and the technology we are discussing here today. When self-driving cars were first introduced, the primitive AI software that drove them was programmed to identify and react to the personalities of human drivers in nearby vehicles. But it often failed. Identifying human personality traits and predicting behavior has gotten a lot better. Now, we are able to add to this the next logical step and transfer that personality to a different vehicle, a new synthetic body that looks just like the person whose personality we transfer.

Today, we ask that you abstain from interfering in this well-functioning market. Transfers will allow people to continue to live and be productive for as long as they want, without illness. This means lower healthcare costs.

We all know now how important the reduction of carbon emissions is. The United States is now getting closer to being carbon neutral. Transfers do not need food. This will mean no net positive production of CO_2. Our technology is good for people; it is good for the planet. It is the market economy working at its best.

We understand that the technology has generated some concerns. Major advances in technology do that. The horse and buggy industry was probably very angry when Henry Ford established the first car company. We plan to address those concerns in a Code of Ethical Practices that we will release within the next month. We think it will address most, if not all, of those concerns.

Simply put, Transfers are another step towards human progress, one that can be compared to Neil Armstrong's first step on the moon. With that, my colleagues and I are happy to answer any questions you may have.

The Chairperson, a Wisconsin New Democrat wearing designer green-rimmed glasses, had publicly expressed serious doubt about transfer technology, so no one on Eidyia's team expected flowers from her.

As soon as Bart was finished, she asked, "Are you comparing yourself to a hero like Neil Armstrong?"

Fortunately, the lobbyist had coached them well for grandstanding. When cameras were rolling, which was almost all the time now, Congressional hearings were not about policy but about scoring political points. The game they played was called gotcha, and it involved creating short clips that could be posted to social media.

"I am not, Chairperson," Bart answered, unfazed. "If there are any heroes here, they are my colleagues who made possible this incredible step forward in technological terms. I only meant that the technological leap we are discussing today is comparable in terms of its impact on our collective psyche to when a human first walked on the moon."

The Chairperson narrowed her eyes, then continued as if reading through a list of prepared questions. "One of the matters we must decide as a committee is whether what you call 'Transfers' are human, and should be treated as such by the law." She looked at Bart, displaying the tough-as-nails look that worked so well in the media. "Would you say that Transfers are human?"

Paul looked at Bart, then pulled the microphone closer when Bart nodded. "If I may, Chairperson. The better question to ask is: What makes people human? It is not any particular body part. Courts have already ruled that enhanced humans, what we used to call cyborgs, are people. Our veterans with synthetic legs, arms, ears, eyes, and other body parts, for example, are not just obviously human; they are, to use the term you just used, heroes. What makes people

human is the mind. And the mind, the personality, is what we transfer."
The training was paying off. Reformulate the question, they'd said.

"But, Gantt-san, what about the soul? Are you saying the soul *transfers*?"

"Good question." Paul smiled, and anyone who knew him would see
that his enigmatic smirk was replete with meaning well beyond an attempt to
project the fake friendliness that's expected of professional witnesses in such
situations. "Biologically, the soul and what we might call the heart, and emotions,
such as people's ability to love, are essentially the result of brain activity. Several
factors influence that activity, including internal and external stimuli, hormones,
and many others. A person's mind reacts to all of this in predictable ways.
By mapping data from the person's behavior using the S-Chip, we can build an
accurate portrait of the individual. And that is what is transferred."

The Chairperson scowled in displeasure. "But is it a transfer or in fact some
mechanical replica of a person? It's a robot, right?"

Many members of Congress were trained as lawyers, and asking leading
questions was the oldest tool in the box. Lawyers, like politicians, used those tools
to score point to win cases, not to get justice. If those funding your reelection
were on the right side of history, so much the better, but then, sometimes they
were not.

"We say Transfer," Paul said. "Because we transfer a personality, and we
only plan to transfer people after their natural death."

Bart was sweating a little, but Paul was as calm as a swan on a smooth pond
on a quiet morning.

The Chairperson frowned. "That is another major concern for this
Committee. I believe the gentleperson from Alabama will have questions for you
later about just that," she said, reaching across the proverbial aisle. "So, are you
saying we should pay social security benefits forever to one of your transfers?"

Paul quickly glanced at the notes he had been given by Eidyia's lobbyists.
"Chairperson, if I may. First, a Transfer will not need health benefits. Then
Transfers do not need to eat or drink, so they can live on a very small budget.
This is also much better for the planet. If the transferor had a family to feed,
the Transfer will be able to work and earn a living."

"I see." The Chairperson's eyes narrowed to two small slits behind her
glasses. "You mean this technology can destroy tens of thousands of American
jobs, replacing workers with Transfers who never get tired?" She moved back
in her chair with that *Touchdown!* look on her face.

Paul was having none of it. "For one thing, the impact of introducing
Transfers into society will be very incremental. As to replacing workers, the real

risk, Chairperson, if you want to look at it that way, is that companies will turn to many types of dumb robots, not Transfers. There are many less expensive options to replace humans."

The Chairperson deflated, like the ref just took her touchdown away based on some arcane rule. She looked at her notes, then turned her chair and spoke to a young person sitting behind her. They exchanged a few whispered words before she turned back to face the Eidyia team with a wry smile on her face.

"Can you give me a straight answer to a simple question? How can you say a Transfer is like a human if there isn't anything human about it? It's a machine, isn't it?"

Paul looked at Koharu.

Their face, like Bart's, was shiny with perspiration. They took the floor, holding their small tablet in their right hand. Their voice was soft when they said, "With all due respect, Chairperson, what is a machine? A machine is something that has various interrelated parts, each of which performs a function, working together to produce some kind of result. That is, in a way, what the body is. Some might say a machine is something made of metal, plastic, or other synthetic material. Humans and other animals have DNA. Some say that makes them alive. If you accept that, then Transfers are not machines. They are made of DNA. It's just a different type of DNA, something called programmable DNA."

The young person behind the Chairperson leaned forward and whispered something. The Chair nodded and asked, "What about consciousness? Does that transfer?"

Koharu's hand was shaking a little under the desk.

Paul looked at their hand and moved the microphone closer to himself. "Consciousness is awareness that one is thinking. 'I think therefore I am,' as Descartes said. Transfers can think. Self-awareness is just another type of thought."

The Chair glanced at the clock in front of her. "Thank you, Gantt-san. Let me turn now to Dr. Annenberg. The gentleperson from Massachusetts, as you know, has a PhD in behavioral genetics."

"Thank you, Chairperson," Annenberg said. He took a sip of water, slowly, deliberately, and looked at Bart for a few seconds. "Van Dick-san, I am trying to understand how you can transfer a person's thinking into a synthetic organism. Human thinking is a hodgepodge of physiological, logical, and other factors. We know that several personality traits are genetic. Others are the result of parenting or other external factors. How can you possibly replicate that?"

Bart took the mic. "Thank you, Dr. Annenberg. That is an excellent question, and that is why we are happy to be here today. To answer questions like that one." He briefly looked in the direction of the Chair, whose eyes shrunk. "In fact, Dr. Annenberg, we made our big step forward when we abandoned attempts to 'replicate' human neural patterns, as artificial intelligence research has tried to do for decades. We realized this was unnecessary and possibly pointless. They were trying to recreate the mechanical vessel, with all the interconnected neurons, hoping that would end up replicating the mind, as you know better than I do. The hope was that some form of consciousness would emerge."

"Yes, of course." The flattery was mollifying Annenberg slowly. His tone became more amiable. He sat up in his high leather chair. "People have discussed this emergence question for a very long time. Are you saying your transfers do not have this emerging consciousness?"

Bart glanced at Koharu. "If I may, I'd like to ask Dr. Tanaka to continue."

Koharu nodded and took the mic. "Thank you." They looked a bit more relaxed after their initial lap. "As Van Dick-san just said, we quickly realized that we needed to replicate the *output*, rather than the *input*. The goal was to transfer whatever outputs a person generated into what I might call a new carrier. As you said, Dr. Annenberg, we know that the heritability of personality traits varies, roughly from thirty to sixty percent. This is in turn due to multiple factors, including genetic predisposition to produce more or less of certain neurotransmitters. This is why replicating the human brain might work for some processes in general, but it's impossible to match a particular person's brain. You could not get that data. So, replicating that as a biological or mechanical matter is essentially impossible in our view."

Bart smiled at them as if to say *well done*.

"So, you don't buy the widely accepted 'nature and nurture' model, Dr. Tanaka?" Annenberg asked, expression incredulous. "Are you telling us today that you believe that is junk science?"

"I apologize if I did not make myself clear, Dr Annenberg. I meant the opposite. It is unquestionably true that most personality traits are the result of some mix of nature and nurture. We also know that many human actions are in fact not the result of conscious thought but nonconscious thought. Think of when you act in a fraction of a second to avoid an accident, or when you walk down the street and your mind is busy doing something else and yet you are walking, avoiding other people and obstacles. The challenge we faced was that the human mind is not…" she paused for effect as the lobbyists had taught them to do, "…rational—and sometimes entirely irrational, it seems—and this is what we wanted to replicate."

Annenberg sat back in his chair, as if a small storm had just passed. After a few seconds, he looked at Koharu again. "What do you mean when you say 'irrational'? All human actions can be explained in some way. They might *seem* irrational. It doesn't mean they *are*."

Paul looked at Koharu, then took the microphone. "I think that actually *explaining* human behavior to a machine is sometimes impossible, Dr Annenberg. Take the way people speak. People use truth—or I should say, things we believe are true—then mix it with half-truths, white lies, and outright lies. But each person tends to do it somewhat differently, depending on the situation."

"I'm not sure I follow." Annenberg frowned.

An audible murmur went through the audience. Were people protesting more or less automatically, which tends to happen whenever someone tells a truth about humans that we don't want to hear? The idea that we're neither fully rational nor truthful, both of which are as obvious as the fact that Tuesday is the day that follows Monday, seems to be one of those shocking admissions, a manner of Victorian prudishness about human nature.

Paul was unfazed, driven by the strength of his deep and unshakeable convictions. "Take someone who is caught cheating on their spouse. They might panic, or they might immediately admit it and beg for an apology, or they might deny it, or they might say 'I can explain' and then use their rational reasoning ability to come up with some half-baked, possibly entirely false account of what happened. That is hard to explain to a logical machine. Add to that that the reason for the affair is itself a source of hard to explain behavioral impulses. Or take someone who smokes. A doctor tells this person they're developing a serious lung disease and need to quit, but they continue to smoke. How do you explain that to a machine? My point is that getting into attempts to replicate the cause of the behavior is not the way to go. What we can replicate is empirical individual patterns of behavior. What we do with our technology is to identify behavior outcomes for each individual. From this, we can *infer* future behavior in essentially any situation. Using our massive datasets, we have identified twenty-seven personality features that explain human behavior. They map reasonably well onto existing psychological knowledge about personality traits, motivations, and so on."

Annenberg shifted in his seat. Was this harder than he'd expected? Arguing with humans meant that almost everyone accepted certain premises by default, but Paul was ignoring those unwritten conventions.

Annenberg tried another angle. "So, people are predictable, is what you're saying? That we are like machines? Or worse, like Pavlovian dogs?" His tone had turned condescending, like he was scolding a bad student.

Paul's inscrutable face didn't reveal even a hint of emotion. It wasn't an image of serenity, more like the blank screen before the movie. He leaned forward again towards the microphone. "Dr Annenberg, my apologies if I wasn't clear." Now he was scolding the teacher. "Twenty-seven different features are actually more than enough to make any of the almost nine billion people on the planet an individual with their own personality. I am not saying that all humans act fully predictably in the sense that everyone will react the same way to the same set of circumstances. I am saying that within a framework of twenty-seven different parameters, you can predict how a specific person will react in any given situation with an extremely high degree of confidence. Humans are different, but they are not infinitely complex."

Annenberg sat back in his chair. "Thank you. That was it for now, Chairperson."

The Chairperson looked at him and then the clock, which still showed three minutes. Annenberg had given up before the bell. "I thank the gentleperson from Massachusetts." Then she turned to the other side. "The gentleperson from Alabama has the floor."

George Corley Thrasher took the floor. His sinewy face was topped by a patch of hair too black to be natural, and a shiny bald spot was visible whenever he looked down at his notes. His dark-rimmed glasses and a suit probably made in the 1990s made for a look from a bygone era. A fervent Evangelical, he was initially seen as a representative of the religious right but his appeal among a broader public was growing.

"Thank you, Chairperson." He nodded and turned to Bart. "Mr. Van Dick, do you believe in God?"

"Well, Representative Thrasher, I am not sure what that has to do with the topic of today's hearing."

"It has EVERYTHING to do with the topic, Mr. Van Dick," Thrasher said, raising his voice like a preacher, his quavering voice amplified by a face quickly changing to the color of hellfire. "These things you call Transfers, they aren't made by God. They are not human!" Twice he pounded his fist on the desk as he spat his venom.

Bart, who had seemed nonplussed up to that point, began moving continuously in his seat as if he had a bee in his pants. Arguing with Evangelicals was probably new to him. Paul looked at him and took the microphone.

"With respect, Representative Thrasher, humans are not made by gods; they are made by other humans. People make babies."

Now, that's a way to close a bible with a snap. A small but audible giggle rippled through the room, and it seemed to deflate Thrasher's balloon. Bart had a hard time covering a smile.

Thrasher's sermonizing voice came down a full octave, but now he was really fuming. "How *dare* you, Mr. Gantt! God made the first humans, and we are just continuing according to His direction."

"I could use the same argument, Representative Thrasher, again with respect. Why did God give us the ability to develop this Transfer technology?"

"Not God, Satan! These Transfers are not humans. They are machines."

"As my colleague Dr. Tanaka explained, the very notion of machine is debatable."

"What isn't debatable, Mr. Gantt, is that your so-called Transfers cannot have faith."

Koharu looked at Paul, who raised an eyebrow in surprise but pushed the mic to them.

"Actually, if I may, Representative Thrasher, if the transferor had faith, then the Transfer will have the same initial belief."

"Belief? Faith is not just some sort of a 'belief.'" He smiled, but it was a smile aimed to kill. He looked around the room. "Saying that chocolate chip cookies are better than pecan pie, that's a belief. Faith comes from God. You just don't get it, do you?"

Koharu's cheeks flushed pink, and Bart stepped in.

"Faith is in the mind. It is a thought. It *is* something we can replicate."

Thrasher drew his mouth back into a sneer, revealing a set of teeth yellowed by thirty years of smoking a pack a day. "You are all godless creatures, just like your so-called Transfers," he spat, his face now fully red and his eyes bulging.

He looked at the Chair and turned his chair away, as if to cast the infidels in front of him to the last circle of Hell. The Chair looked at him and then turned towards the other side of the podium.

"Our last questions for this session will be from the gentleperson from Oregon."

Sophie Loren took the floor, poised as always. "Thank you, Chairperson."

This was Loren's second term in Congress. She had demonstrated many times an ability to work across the aisle, or, in her case, both aisles. She was one of the rare third-party candidates in the history of the House of Representatives, a member of the Social Libertarians, a party that advocated a "strong but intelligent" safety net and reduced government interference in people's lives. Almost forty percent of voters under thirty had voted Social Libertarian in the last federal elections. Always dressed impeccably, like she belonged on the cover of a business magazine, Loren was highly popular and played political games with brio and flair.

She had a look of genuine concern on her face when she said, "I am not sure whom to ask this of, frankly, but my question is this: Are Transfers basically machines that do as programmed, or do they have free will?"

Paul looked at Bart and Koharu. They both nodded, and he took the floor.

"Thank you very much, Representative Loren. It will not surprise you that I want to pick up where my colleague Dr. Tanaka ended earlier. People are, in a way, predictable. It's just that each human being is, well, predictably different. I don't mean this in a mechanistic or Newtonian way. I mean that some human actions are not the product of deliberate thought, like people reacting quickly to a sudden event. People do have some degree of agency, of course. They decide certain things after a careful and deliberate reflection phase, but that is not entirely free will. After all, the largest industry in the world right now is advertising. Even members of this House spend most of their campaign budget on advertising."

The lobbyists had told them that Loren spent less than others on traditional advertising, counting instead on a large contingent of followers on social media. She was starting to nod in agreement.

Paul seized on his momentum. "People can be influenced quite easily by those external factors, and other times by what I might call internal ones, like neuroses or some other non-conscious factor."

Loren's nod stopped and she tilted her head to the side. "So, you admit that Transfers have no free will, but that is because you think humans have no free will?"

"Not at all. People definitely have *some degree* of free will," Paul said, pronouncing the words slowly. "But when you measure all that a person does, including the decisions that person makes, even knowing that some of those decisions are at least in part the result of what we might call 'free will,' you can accurately predict what that same person will do in the future when faced with other decisions."

Loren took a moment to digest this information. "So, if I can take an example, you mean that if someone was offered a new job, say, in a different part of the country, you could predict that person's answer?"

"Yes, with high accuracy. And what we transfer is the data, the life experience if you will, that makes the decisions. So, the Transfer will adapt to the new job and the new city and state the same way the transferor would have."

The Chair glanced at the clock and turned towards Loren. "Thank you, Representative Loren." She looked at Bart, Koharu, and Paul with a hard face. "Thank you to the three of you for your time today," she said without a hint of sincerity. "This afternoon we will hear from Professor Sam Reidenberg, Chair of Robot Ethics at Harvard University, and Professor Christine Jacobs from Knights University."

CHAPTER 22

At two p.m., the Chairperson reopened the hearing with her gavel. Christine and Sam Reidenberg were sworn in.

"Dr. Jacobs," the Chair said, "let me start with you. I understand that you work for Eidyia, is that right?"

Christine, wearing her mother's pearl necklace over a white blouse and navy-blue pantsuit, moved closer to the microphone. Her heart was beating much faster than it should. "Thank you, Chairperson. I am a Full Professor at Knights University Law School." Christine could hear the sharp pitch of her own voice. *Calm down*. She took a deep breath.

"You mean the...*Van*...*Dick* Law School?" the Chair asked. "As in, the Mister Van Dick we heard from this morning?"

"Yes, Chairperson. That is the official name."

"Soooo, the school was named after Bart Van Dick, from Eidyia, is that correct?"

"I believe so." *Not good*, Christine thought. *Stick to 'yes' when it's obvious*.

"So, you *do* work for Eidyia after all? Why not come out and say it? We expect straight answers from you, Professor."

She'd expected a grilling, but this was not an auspicious start. *I need to dig out of this hole*. Then she noticed, sitting on a bench against the wall behind the members of Congress, a young woman with freckles and long, straight black hair she thought recognized. She was holding a tablet but looking at the room. What was her name again? Daphne Desmarais! Yes, she'd been Christine's research assistant a few years ago. She looked at Daphne and smiled with a very gentle nod, so that no one else would notice. Daphne smiled back. That brief contact recharged her batteries.

"With respect, Chairperson. Hundreds of law schools, business schools, libraries, and in fact some entire universities are named after donors. This is part of why the United States is the world leader in higher education. We have resources that other countries don't have. In all cases, or at least all cases that I am familiar with, the donors have little if any influence on what goes on inside the school or university. Professors have academic freedom." Would that do?

"But," the Chair countered, "you would not write anything that Eidyia might disagree with, would you?"

"I would not hesitate to do so, Chairperson," Christine said, perhaps too confidently.

"Okay. Can you point us to anything you've said or written that Eidyia disagrees with?"

Oh, she had fallen into a trap. Then a way out appeared. "You would have to ask Eidyia."

"Hmm, well."

Christine felt she had come out even, but the Chairwoman wasn't finished.

"You do work for Eidyia directly on this Transfer Project, don't you?"

"Eidyia asked me for an opinion on certain legal questions concerning the project."

"Were you paid?"

"They paid for my travel costs when I was asked to join them for a meeting."

"I see. And you also work for the Department of Defense?"

"Again, I would not say I 'work for' them, Chairperson. I was asked to attend a meeting and to be part of the US delegation at a UN meeting in Geneva in an advisory capacity."

"That's the conference on the use of robots in wars and warlike situations?"

"Yes. Correct." *Keep rolling with the punches*, she told herself. She could score her own points with questions from other members later.

"I see. Can you tell us what legal questions Eidyia asked you to opine on?"

"With pleasure." She was glad they had discussed this at lunch. "First, there was the question of having two versions of the same person. I mean, if you had a person and added a Transfer that was virtually indistinguishable from that person."

"And what advice did you give them?"

"That, as a legal matter, there is no obvious answer. Arguably, you have two, well, agents, as we say in law, making decisions that may be legally binding."

"I went to law school, Dr. Jacobs. I know what an agent is." The Chair gave her a snarky smile. "What I don't understand is, why would that matter? We heard this morning that a Transfer would make exactly the same decisions as the, what did your colleagues call it, transferor? So, this was not correct, I take it? Were we misled?"

"As far as I know that statement from Eidyia was entirely correct, but perhaps there is more to the question. A decision is the product in part of external factors. The same person may even make contradictory decisions, and two versions of the same person living different experiences are likely to do so at least some of the time. Unless each version knew exactly what the other version had decided in real-time, it would cause difficult situations."

"So let me you ask again, Dr. Jacobs, this morning when the Eidyia team said they could predict decisions with high accuracy, were they misleading us?"

"No, Chairperson. What I am saying is that you need to know all external factors that might influence the person."

"Such as?"

"Well, let me give you an example." Now that she'd found her teacher's stride, Christine was unflappable. "Assume one version of A, say the transferor, makes a contract with person B. Then B meets the other version of A, the Transfer, who does not know about the contract. That would cause legal issues and create potentially embarrassing social situations as well."

"And what was your advice?"

"I advised against having two versions in existence at the same time. Lawyers are trained to minimize or mitigate risks." She paused. "As you know," she added with a smile. "This is why, I believe, Eidyia chose to allow transfers only at the time of death."

The Chairperson looked at the young person behind her, who pursed their lips and tilted their head, signaling uncertainty. The Chair turned back to face the room. "I see. Thank you, Dr. Jacobs." She looked to Annenberg. "The gentleperson from Massachusetts has the floor."

Christine had been keeping mental score, and she thought the game was tied. How could they get ahead?

"Thank you, Chairperson," Annenberg said. "Dr. Jacobs, in your view, are the so-called Transfers human?"

"As a matter of law, that is a difficult question, Dr. Annenberg. Biologically, they are made of DNA that is programmed to function almost like human DNA. Humans and some animals share a very high percentage of DNA, as you know better than I do."

"Yes, more or less ninety-eight percent. The question is, is DNA what makes someone human? Wouldn't you agree? Eidyia uses *synthetic* DNA, correct?"

"As matter of law, I am not sure DNA is the deciding factor. Humans have considerable DNA variability."

"Dr. Jacobs, unlike so many of my colleagues here, I'm not a lawyer. I'm just trying to understand."

Self-deprecation. Figures. Be careful.

Annenberg gave her a challenging stare. "Tell me, are all humans equal under the Constitution, whether or not there are variations in their DNA?"

"Under the Constitution, yes, of course," Christine said, smiling but feeling the ground she was on getting slippery. *Tread care-ful-ly.* "To be very clear, I was obviously not referring to the form of segregation that was authorized under the Constitution until the Fourteenth and Fifteenth Amendments. What I meant was that not just humans but legal persons such as corporations have constitutional rights. Children or people in a coma, for example, have rights too, but a different legal status. People between the ages of eighteen and twenty-one can vote and be called on to defend their country but cannot buy alcohol. And the list goes on."

Perhaps Annenberg's disappointing bout with Paul in the morning had increased his motivation. One way or the other, he looked like a dog not ready to let go of the bone. "You have not answered my question, Dr Jacobs. Are Transfers humans? I mean as a matter of law."

Christine was used to tough arguments. Law students could be tougher than anyone. But this "doctor to doctor" exchange made her feel more confident. She looked briefly at Daphne, who was busy taking notes.

"I was getting there, Dr. Annenberg. I think what makes human beings human as a matter of law is a special type of agency. The law regards decisions we make as binding because we make them in what we like to think of as a freeway. In that sense, a Transfer is like a human."

"Thank you, time is up," the Chair said suddenly, interrupting what had begun to sound like a conversation.

A draw at best, Christine thought.

The Chair turned to her right. "We now turn to the gentleperson from Alabama."

"Thank you, Chairperson," Thrasher said. He looked menacingly around the room and then his gaze landed on Christine, who met his glare with apprehension. If human eyes could throw flames… "Soooo, you are telling us today, Dr. Jacobs, that robots should have equal protection under the law?"

Christine pretended to look at her tablet briefly. She took a quick sip of water, a tip from the lobbyist. He had called it "your time to think." She returned Thrasher's gaze. *You probably didn't do too well in a courtroom, buddy. That was a lame leading question.* "No, I am not saying that at all, Representative Thrasher. I am saying that Transfers are *like* humans in some respects because they function like humans. Most robots do not."

"Ahh!" Thrasher leaned back in his seat and steepled his fingers. "So, you mean to say that Transfers aren't robots?"

"Yes. I do."

"How can you say that? They are. Everyone knows that. Anyone can see that."

Christine blinked to keep from rolling her eyes. *Ah, the old "everyone knows" trick. Up there with "I heard…" Doesn't cut it, pal.* "In my view, robots are machines that can make decisions and sometimes replicate human functions. They are often built to perform repetitive tasks. A robot does not have self-awareness."

"Ah ha!" Thrasher said. "Finally, so you *agree* that it is our God-given gift, our conscience that makes us human! We make decisions every day based on what our conscience dictates."

I wish. She took another sip. *Careful now.* She looked at her notes and then at Thrasher. "Yes, I would tend to agree with that statement, Representative Thrasher."

Thrasher smiled as if he had just won the match.

But Christine wasn't done. She had a bag of tricks of her own. Agree first, then disagree and nail it. "The phenomenon we call conscience, or self-awareness, is not just about making decisions. Animals make decisions and even foresee some of the consequences of their decisions. It is the ability to reflect upon our existence and factor in moral and other considerations that makes us human."

Thrasher's face changed, expression turning perplexed. "Moral considerations? You mean faith, right?"

This was like talking to a student. "Faith can be the source of that for some people. Ethics can be derived from many other sources, of course."

"Not real ones!" Thrasher's face was turning red again.

The Chair cut in. "Two minutes."

He looked surprised that his time was almost up. After a glance at his notes, he said, "Okay. Last question. For now. Dr. Jacobs, if a Transfer and its, how did they call it this morning (he pretends to look at his notes), transferor existed at the same time, to use the example you mentioned, and that person was married, then as you see it, the spouse would have two husbands or two wives, right?"

"I must say I never considered that angle, Representative Thrasher." She caught a glimpse of Bart smiling in the row behind her.

"Would the spouse be committing adultery if he or she was with the Transfer?"

"I…I must say I don't know, but that might be another good reason not to have two versions of the same person." Oh, he had scored one there.

"And the transferor could ask for divorce due to the *adultery*, right?" He said the word with fire on his tongue, like a bible just hit the table.

"Again, sir, with respect, I didn't give those questions any thought because I was of the view that there should be only one version in existence at any given time." *Come on, Christine, you can do better.*

"Well, then, Dr. Jacobs, explain this to me. If it is adultery if the transferor is still alive, why is it not when he is dead?" Thrasher looked triumphant.

"Perhaps because you cannot commit adultery if the spouse is no longer alive? Till death do us part, I believe."

Had she evened the score?

The Chairperson looked at Thrasher. "I thank the gentleperson from Alabama." There was a smile on her face for the first time since the beginning of the hearing. "I give the floor to the gentleperson from Oregon."

"Thank you, Chairperson." Loren took a sip from the paper cup in front of her.

I wonder if that's West Coast coffee. I could use one right now.

Loren looked at her. "Dr. Jacobs, let me pick up on something you said earlier. You said 'we like to think it is free will' about people making decisions. Do you mean we are not free to think as we please?"

Christine, who had taken a moment to lean back and look at Bart and Paul, moved closer to the mike. "Thank you for your question, Representative Loren. As a matter of law, there are of course situations we call coercion, like when someone asks you to sign a contract with a gun to the head. Otherwise, when we have time to think, the law usually regards our actions as having legal significance. A contract is binding even if the salesperson was using psychological techniques to convince you to sign, like loss aversion." Mechanically, as if to mimic Loren, Christine took a sip of water.

"Loss aversion? I guess I missed that one in college," Loren said with a smile.

"Say you offer someone a deal for a limited period, but the urgency is entirely made up. That creates a desire not to lose the opportunity."

"I see. Okay. Please continue."

"I said we *think* we are free to think," Christine said, pronouncing the word like she was dropping a rock. "I was echoing what Gantt-san said this morning about free will. We like to think we have free will, but the reality is that our actions are the product of several factors, many of which are non-conscious."

"I am familiar with these ideas that we have limited free will. I won't debate that with you today." Loren took another sip. "Let me move on. In your view, as I understand it, Transfers are human?"

"I would say they are persons. They are *like* humans." Christine ran through her discussion with the lobbyist about how to work with Loren's libertarian views. "I, for one, would give anyone the freedom to transfer their personality, their consciousness if you will, in a different carrier, to quote Dr. Tanaka." Christine hoped this would hit the target. It looked like it may have.

"I'll have to give that more thought." Loren looked at her tablet. "Tell me, Dr. Jacobs, do Transfers dream?"

"Dream? Hmm, that's a good question!" Christine smiled mechanically as her brain changed gears.

Loren pushed on. "Don't you think dreaming is part of what makes us human?"

"Alexander Pushkin, probably the most famous Russian poet. Nineteenth century. Not my favorite, though. Anyways, he once wrote something like 'it is better to have dreamed a thousand dreams that never were than never to have dreamed at all.' I think he meant dreaming as in having a vision of the future, and I think your question is about actual dreaming, REM sleep and so on. I am no expert in that field, but I am not aware of any legal reason to think of dreams as necessary to be human. I assume most people dream; they are an important information-processing tool. I would say the question is not legally relevant."

Loren looked like she had no idea what to say next.

The Chairperson bent forward and looked at her. "I thank the gentleperson from Oregon. And thank you, Dr. Jacobs," she said, turning to Christine.

And with that, a short recess was called.

***

"Professor Reidenberg, thank you for coming today." Christine noticed that the Chairperson had adopted a warm and friendly tone for the first time today. "We have a few questions about Eidyia's so-called 'transfer technology.' I understand that you have some familiarity with this technology."

Reidenberg looked small sitting next to Christine. They were probably five-foot-four and moved constantly in their seat, their hair already almost all white with a few strands on top that looked like a creative mess. "Thank you, Chairperson. Yes, some. I read the press reports and Eidyia's briefing book. I was also here this morning to listen to the testimony given by Eidyia representatives."

"Very well then. As you know, we are trying to decide whether Transfers should have rights as humans. What is your view?"

The lobbyists had told Christine to look for any hooks she could use if given more time later, so she listened intently.

"My view, is that no, they are not human, but they should have some rights."

Relief rushed over her. She had hoped but wasn't sure Reidenberg would take this middle road.

The Chair did not seem pleased. "Could you elaborate? What kind of rights, and why?"

"Yes, of course. My view is that what makes humans is not DNA or even self-awareness. I am not sure we can prove that there are no animals with some sort of self-awareness. Or robots for that matter."

Tsk tsk. Double negative.

Reidenberg went on. "Self-awareness may actually exist in different forms. I would add to this that we are the only species that is aware of our own mortality."

"But, Dr Reidenberg, we heard from Eidyia this morning that all kinds of thoughts can be, I guess, *replicated* in Transfers? Why not this one, this, hmmm, awareness?"

Reidenberg ran their hand through the untamed white strands on their head. Then they leaned forward. "Precisely because Transfers do not die. When part of a Transfer is damaged, it can be repaired or replaced. In the worst-case scenario, it can be re-transferred to a new body. In theory, Transfers can live forever."

Christine tried not to smile. Reidenberg was turning into an Eidyia's sales rep. And they weren't done.

"I mean, humans generally want to live as long as possible, but we always knew, as we know now, that there is an end, whether it comes at age twenty or one hundred and twenty. Without that knowledge, an organism may be alive, it may even have self-awareness, but it is not human."

Christine deflated a bit. *Oh, that's not so good.* Her watch buzzed, reminding her for the nth time to have her S-Chip checked.

The Chair seemed to pick up on her disappearing smile. "I see. Interesting." She looked to the side. "Thank you. The gentleperson from Massachusetts."

"Thank you, Chairperson," Annenberg said, sitting up confidently. He probably expected Reidenberg to be an ally. "Dr. Reidenberg, you said Transfers are not human but should have some rights. I don't understand why. What did you mean?"

"Thank you for the question, Dr. Annenberg. I meant that there are non-human, let's say, entities that should have some sort of rights. Animals have rights. For example, you cannot torture your dog. I believe this is what the law says. We give corporations rights that are very close to the same rights that people have. People in Ohio gave rights to a lake back in the 2010s."

"So, you think there is a case for giving these robots, these so-called Transfers, some sort of rights?"

"I understand from the briefing book that Transfers can feel something like pain and pleasure. I would call that sentience. If that is the case, then we could recognize that humans should not cause them unnecessary pain."

"You mean, like animals?"

"Yes, something like that."

Annenberg frowned. "I see. Thank you, I yield the rest of my time, Chairperson."

Surprise flashed across the Chairperson's face. "I thank the gentleperson from Massachusetts." She gave him a strange look then turned to the other side. "The gentleperson from Alabama."

"Thank you, Chairperson," Thrasher said. "Dr. Reidenberg, you said humans value life because it ends. You're not a person of faith then. You don't believe in the afterlife?"

A flash of irritation crossed Reidenberg's face. "Representative Thrasher, I prefer not to discuss my personal beliefs. What I can say is that faith and belief in life after death obviously exist. Some say those beliefs have been generated by evolution," Thrasher's lips curled up in disgust, "and by our knowledge that life ends. It is, in a way, a coping mechanism. A creature without a sense of its own mortality, or one who has no reason to think that life will end someday, would not need those beliefs."

Thrasher leaned forward and pointed at Reidenberg. "That sounds like the nonsense I hear from atheists, Doctor. Do you mean to say in front of all these people here that life after death isn't something we can only be given by God?"

Another old trick. Try to get the person to feel bad because of others. It seemed to affect Reidenberg a bit. "I did not mean it that way, Representative Thrasher. I meant that without death, religion might exist in a very different way. I also know this: One reason we can say with certainty that God is not human is that God does not die."

"I see, but at least you agree that these transfer machines aren't human. If they don't die, if I follow your reasoning, they won't have our religion?"

"I am not sure what you mean when you say '*our* religion,' Representative Thrasher. I find it odd to think that if a human being's thoughts were transferred in an organism that doesn't die, that the person's faith or belief in life after death would make much sense at all."

"Our time is up for today," the Chairperson interjected. "On behalf of all my colleagues, let me thank Dr. Reidenberg." She paused. "And Dr Jacobs too."

CHAPTER 23

Back at the boutique hotel where Paul always booked a room when he stayed at in DC, he and Christine opened a bottle of Pouilly-Fuissé.

"Do you like the wine?" Paul asked as he took a sip. "The French make good Chardonnay. And cheers, by the way! Well done today."

"Thank you!" She was happy to be done, but something about the day lingered in her mind. She looked at Paul, who was staring at his glass pensively.

"I wanted to ask you, you know what this Reidenberg said today, about mortality and all that. What do you make of it?"

That pushed her strange thought to the back of her mind, wherever that was. "He took a page, or a frame, straight out of Tarkovsky. In *Solaris* they say that the unique self-awareness of humans lies in the fact that of the billions of creatures on this planet, we are the only ones who are aware of our own mortality, which gives humans a sense of finality that other creatures do not have."

"Damn! Tarkovsky should have been there today!"

Christine smiled. "That would have been fun! He might have had a difficult time with Representative Thrasher, though."

Paul sneered at the mention of his name. "Why?"

"Well, the discussion in *Solaris* suggests that humans invented religion to deal with the anxiety we feel about our inevitable finality. One of the characters says something like 'religion has been bred into us.' That it is not *why* we have a sense of our own mortality; it is the consequence." With her glass in her hand, she lay down on the king size bed covered with a pale blue bedspread. From the mountain of pillows of various shapes and sizes, she picked up a few and made an improvised backrest against the upholstered headboard. She pushed off her shoes with her feet and took a sip.

Paul was looking at her. "Inevitable, until now! *Solaris* was so close to today's reality."

Christine swished her glass like Paul had taught her. "Yes, as sci-fi often is. Great movies are closer to life than life itself sometimes. You simply cannot watch the Russian version and think of the Mosfilm studios where it was made, or of soundstage twenty at Warner Brothers' studios while watching the US version. You are on Solaris with them, not in Moscow or Burbank. The fact that this sounds so true explains why so many church people have come out against the technology. They might be out of business!"

"I was a bit surprised when you brought up Pushkin today. What was the point? And besides, that was risky."

"Why?"

"They might think you're a closet communist or something!"

"Seriously?" She frowned. "I like Russian poetry and Tarkovsky movies, yes, but I also like a lot of other filmmakers who can put poetry into moving images. Anyway, I'm not sure what I meant, maybe that dreaming can mean a lot of different things." Christine got up and opened the minibar. She took a bag of chips and a small can of Greek olives out and went back to the bed, where she rearranged the pillows again.

"These chips are good," she teased. "Pity, you can't enjoy them."

Paul pouted, almost in disgust it seemed to her.

"You know, Paul, most of the time, I wonder like you seem to if, as a species, humans are actually making progress. We almost destroyed the planet to put paper wealth in the hands of the few. The music and film industries are now AI-fueled cash-generating machines instead of being some of the best ways in which human beings can express their creativity. This is what happens with capitalism."

"So, you *are* a communist after all!"

"That's just a label." She took a sip and looked out the window at the Washington Monument in the distance. "Why would you say that?"

"I'm kidding," he snickered. "You just don't sound like a very convincing capitalist."

Christine grinned. She knew, and knew that Paul knew, that anyone under forty was aware not only that there were many widely diverging forms of capitalism, as the American and Chinese models had shown, but also that, contrary to the belief held by the generations that had lived through the collapse of the Soviet bloc, capitalism was not by any means some natural state that had the right to exist forever. In two centuries, it had created more inequality and destroyed more ecosystems than the thousands of years of human activity that came before.

Paul refilled her glass. "I'd better call downstairs. We'll need a second one."

She opened the can of olives and counted them. "Seven olives for what? Twenty dollars?"

Paul got up, walked to the minibar, and picked up the list. "Twenty-two actually. See, Chrissie, capitalism does work!"

She gave him the look. "I'll go shower."

When she was done, she got into bed. Paul was typing on his laptop. She could really have used a sleeping companion, and more, right now. Was it the wine? She was getting increasingly comfortable with Paul Prime, often forgetting he wasn't human, but then her conflicting thoughts about Paul's ruse always closed the door that her desires started to open. And it wasn't like this version of Paul was trying to open it.

"Since I don't need to sleep, I'll just keep plugging away at the protocol for the live tests," he said without looking up.

Christine turned to her side and pulled the covers over her shoulder. With the wine and melatonin running through her veins, she was soon fast asleep.

CHAPTER 24

When she got home the next day, Christine finally got her S-Chip replaced at a local clinic. She was in a let's sort this out mood, like when you sit down to pay all your overdue bills. When the new chip was activated, it detected an "anomaly," so she made an appointment to see Dr. Patel the next morning.

As soon as the doctor entered the exam room, Christine could tell something was wrong. It was the same Dr. Patel, but with a different face. He gave her the news, and it was like a punch in the stomach. She could feel her guts shaking. Dr. Patel gave her a small injection, and her body calmed down a bit. She left the hospital sleepwalking and got into a PC.

When she got home, she called Rachel three times. The third time, she left a voicemail. Then she called Paul. He answered on the first ring, and without preamble, she said, "Paul, come quick. I need you. Like, now. Please."

"Of course," he agreed without hesitation, concern lacing his voice. "I can be there in a few hours. I will see you then." Perks of having a private jet.

Harry gave her a shot of the opioids prescribed by Dr. Patel, and Christine fell into an unsettled sleep. She awoke to the sound of knocking at the front door and pulled herself off the couch groggily. The minute she saw Paul on her doorstep, she fell into his arms, sobbing.

"Paul, I have cancer! Fucking cancer! Like my mom!"

Paul was silent for a minute. "Cancer? Oh, Chrissie. I'm so sorry. But they catch it so early these days with the S-Chip that it shouldn't be a big problem, right?"

She sobbed harder and put her head on his shoulder. "It's my fault, Paul. My S-Chip was malfunctioning for months, and it took me forever to get it replaced."

"Oh." He hesitated. "So, what did the doctors say?" He took a small step to move them inside the townhouse, then closed the door behind them.

"Not good, Paul."

He held her tighter, rubbing her back. Part of her mind still knew this was Paul Prime, but she no longer had the energy or even the will to resist. She needed Paul, and this was close enough.

After long minutes of uncontrolled crying, Christine tried to get her jumbled thoughts into some sort of order. She dug deep to find an ounce of calm somewhere inside her. "There is a treatment, but the doctor says the chances it will work are less than fifty percent. Much less. And there are terrible side effects."

"Let's have a glass of wine and talk about it." Paul turned to the robot. "Harry, is there any wine? Bring us a bottle of red."

A bottle and a half later, Christine was cried out and felt more relaxed.

Paul put his glass down and took her hands in his. "Chrissie, have you considered avoiding entirely that dark tunnel in front of you? The side effects, the sleepless nights, the anxiety, and the chance that you will go through it all for nothing?"

"I…I'm not sure I follow, Paul."

"What I'm saying is that you could think about a transfer."

She looked at him with eyes the size of half-dollars. "Are you fucking kidding me, Paul? You want me to *die*?"

He took her hands in his and looked straight in her eyes. "No, Chrissie, I want you to *live*. Forever." He looked down for a second and then reconnected with her wounded gaze. "Look, every human is going to die someday. In your case, it may be that it will come sooner than expected and only after a good deal of pain and suffering."

"Oh, Paul. You cannot be serious." Another wave of sobs overwhelmed her. Not cried out after all.

Paul stayed silent, soothingly rubbing her hands. When her breathing was more under control, he said, "I'm dead serious, Christine."

Christine tried to fight the buzz on which her neurons were slipping as they were trying to grip what Paul was saying. Her emotions were tangled up in an unfathomable knot. She felt like she had just taken a dive in deep waters, where all is at once eerily quiet and deeply unnerving. She asked Harry to bring her a large glass of ice water and gulped it down.

They remained silent for a few minutes. She was floating in fog of deepening torment. Between the wine and the opioids, she barely felt anything when Paul picked up her limp body and brought her to bed, undressed her, and covered her with the thick-down comforter. Dewey jumped up and curled up on the pillow next to her. She didn't see that she had missed calls from Rachel until the next morning.

When she woke up, Paul had already left. Harry delivered a voice message saying he had to leave but would be back as soon as possible. Fighting a searing headache, she drank several cups of strong, dark coffee, managed to shower, then started combing her hair. Her eyes were still red and swollen. Seeing yourself like that in the mirror is enough to send you back for another loop on a path of self-imposed misery. She moved to the couch and asked Harry to bring her another cup of coffee. Dewey was never more than a step away from her. He just walked around her softly, like a nun. Cats know.

She called Rachel and started sobbing as soon as she picked up. It took several long minutes of wiping her eyes and her runny nose with her housecoat to get enough of a grip to tell Rachel what was going on. Rachel offered to come see her the next day, but it was only a few days until Christine's planned trip to Nashville, and she insisted she'd be okay until then.

CHAPTER 25

At the newly expanded Nashville airport, many people were wearing cowboy boots and hats. They looked strangely out of place in a city now largely populated by people from the Northeast or the West Coast, including 40,000 mostly conservative Californians and New Yorkers who had moved there in the past three years alone. Christine took a PC to Rachel's condo, remembering the first time she'd walked on Broadway with its noisy honky-tonks—now full of holographic versions of old country music greats.

The PC stopped in front of Rachel's gleaming building near the live music bars on Demonbreun. Christine punched the code on the small guest video screen next to the entrance. Rachel was waiting in the open doorway when she reached her floor.

"Hi, Rachel. It's so good to see you!" Christine was making herculean efforts to keep it together, but she could feel her insides melting.

They hugged for a long time.

"Come in!" was all that Rachel said, instinctively knowing, in the way that best friends do, that opening the tap would break a huge dam.

As she walked in, Christine was struck by the odor of fresh bread. The smell filled her lungs, bringing back memories of Rachel, so proud of her old sourdough starter, trying to teach her how to knead. Christine remembered massaging the soft, slightly sourdough until it no longer stuck to her hands, with Rachel often standing behind her.

"Still making that famous sourdough bread of yours?" Christine asked.

"Yes! The starter is even better now. Almost twenty-two years old. I made a rye loaf for you—your favorite."

"I can't wait!" she said, forcing a smile. She *needed* to talk to Rachel, and tell her *everything*, until her entire bag of jumbled emotions was emptied, but where would she begin?

"Put your stuff in the guest bedroom," Rachel said. "I'll get changed and we can go for a walk. It's so nice out today. First day below eighty degrees since May, I think!"

"What? It's nearly December!"

"Yeah. We broke records again. No rain for three months, then two days above one-twenty in July, and then two hurricanes in September. Not as bad as last year though."

Christine brought her bag to the small bedroom across from Rachel's owner's suite. She recognized the queen bed with the quilted duvet cover Rachel had purchased in Vermont when they were on a skiing trip in Stowe. The prints on the wall hadn't changed either. A Toulouse-Lautrec print from the Grand-Palais in Paris. Had Rachel even been there? And a picture of her father's sailboat, with a proud nine-year-old Rachel holding the big wheel and a smiling, shirtless, burly man wearing one of those captain's hat you can buy at a souvenir shop near a pier standing behind her.

Christine took a deep breath and went back to the living room. "I'm ready!"

Rachel had her hands deep in that kitchen drawer where a thousand little nothings end up. "Can't find my keys." Then she held up the key ring with a miniature shiny red guitar hanging from it like a trophy. "Let's go!"

As they left the building, Christine noticed the solar panels covering almost every rooftop. There had been debates in the early 2020s when Tennessee had imposed a tax to discourage people from installing them. It seemed things had finally changed for the better.

"Where do you want to go?" Rachel asked.

"I'm just happy to walk and chat. Anywhere is good." This was truly a *not the destination, just the road there* moment. Her hands were trembling slightly.

"If we go right, we'll be on what people still call Music Row, even though almost all the music stuff is gone."

"Isn't that your fault?" Christine asked with a faint smile.

"Don't joke about that! Some days I feel we have lost so much. It's the first time in, what, thousands of years, that humans are basically not creating new music? All done by machines."

Without thinking about it, at least not consciously, they started walking toward the Vanderbilt campus. Signs proudly announcing its "national arboretum" status looked a bit out of place with so many trees felled by extreme weather events in recent years. Still, it was nice to slip back into their lives as twenty-somethings.

Neither of them said anything for a while, but as they walked towards the center of campus, passing the high bell tower of Kirkland Hall, Rachel grabbed Christine's hand.

"Just like old times, sis."

They crossed the bridge over 21st Avenue to the Commons, as they had done a million times before, and after another short climb, arrived in front of East House, their freshman dorm. The old building had been renovated, and not much of the original spirit seemed to remain. A student was walking out from a side entrance as they approached.

"Hey, hi, sorry," Rachel called out. "Could you keep that door open for a sec? We spent a year in this building and just wanted to have quick peek at the hands on the wall."

"Sure," the kid said, eyeing them with mild suspicion.

In the basement, students had been leaving handprints on the wall with their names and year for decades. It took a while, but eventually they found their own. Christine's purple handprint was still visible, though partially covered by a more recent hand added ten years later. Rachel's was also there, right next to hers.

"Good times," Rachel said. "Remember when that guy, what was his name, the tall one with curly hair who played basketball?"

"Harry. Like my robot."

"Yeah, right, Harry. He came down here at like two in the morning, and found me and that girl I was dating half naked on the big couch?"

"I do!" Christine laughed nervously. Her mind was going in a million directions, but she tried to sink into Rachel's soothing presence. Sister therapy, she called it. She could always turn to Rachel and be herself. No questions asked. Like in the old days.

Rachel trailed her fingers over her old handprint, then stepped away from the wall. "Say, I'm getting hungry. Do you still like hot chicken?"

That pulled Christine right out of her daydream. She wasn't hungry, but if Rachel was, they should eat. "Sure."

They walked down Division to the original Hattie B's. Now that there were something like twelve Hattie B's in Nashville it was much easier to get in and they wouldn't have to line up for an hour. The original location was ensconced under a huge residential tower, making the old restaurant look like a recalcitrant homeowner who won't move to allow a new highway to be built. They ordered from the tablets on the table.

Christine knew she'd have to talk about the cancer and the possibility of transferring soon, but she didn't want to get there. Not just yet. "Tell me, what's going on in your life? How are things with Matt?"

"Well, you know it's been on and off for almost three years now. He's nice and kind. Also bi. But I just don't think it's going anywhere. He's my go-to guy, but kind of in-between, you know? On Music Row, people just go from lover to lover. I feel like a bird picking from one flower after the next sometimes."

Even though she'd been the one to bring it up, Christine quickly realized she wasn't in the mood to talk about Matt, or Rachel's love life. She had too much on her mind. Luckily, two steaming baskets of chicken and fries landed on their table just then.

Rachel took the first bite, looking at Christine. "Man!"

Christine moved the fries covering her chicken and picked up a wing from the basket in front of her. She took a bite and immediately followed it with a sip of water. "Yeah, it's way spicier than I remembered. I must be getting old."

They were both crying by the time they finished eating, but it was strangely therapeutic. Christine's emotional hurricane had turned into a tropical storm, matched by the tempest on her taste buds. Strong enough to flood.

They went back to Rachel's apartment still feeling the burn in their mouths. Christine sat on the big, old, sky-blue fake velour couch Rachel had had for years.

"Chardonnay?" Rachel asked.

"Sure, that would be nice. Thanks."

Rachel poured them each a very large glass and then joined Christine on the couch after putting the bottle in a bucket of ice water on the coffee table. After all that spice, the heavily oaked Sonoma white tasted like water. Christine stared down at her hands. She really should stop biting her nails.

Rachel studied her. Christine was still looking at her hands but not seeing them anymore. Her vision blurred. A few tears at first and then the floodgates just opened.

Rachel moved closer and put her arms around her. "Tell me. Everything."

Just what Christine needed to do. What sisters are for. After a few minutes, she regained enough composure to speak, but her nose was stuffed and her eyes swollen. "As I told you on the phone, I'm...I'm sick. I mean very."

Rachel drank almost half her glass in one sip, as if looking for appropriate words in the buttery wine. "Can I ask, don't these things get, you know, detected these days?"

"Yeah, they should, but my S-Chip malfunctioned, and now it's basically..." She hesitated. "Too late." Another groundswell of sobs overwhelmed her.

Rachel sat with her in uncharacteristic silence. But that was okay; all Christine needed in that moment was a shoulder to cry on.

"But there is a way out," she finally said.

Rachel looked up from behind her glass. "What, an experimental treatment?"

Christine stared into her wine. "No. Well, not really."

Rachel rolled her head back in surprise. "You don't mean assisted suicide, do you?"

Christine tried to smile. She raised her head to look at Rachel. "No. I mean, again, not really."

Rachel put her glass down, scooted closer to Christine, and turned towards her. "Not really? What does that even *mean*? Christine, tell me!"

Christine felt her friend's growing concern and desperation for answers. "Have you heard about the Transfer Project?"

Rachel nodded. "Of course. It's all they're talking about on the news."

"People have started to transfer. Tests, you know. And it's working well."

"Wait. Don't tell me you're thinking about—"

"Paul has transferred. No one knows, so please keep it to yourself." In her normal state, Christine would never have revealed that, but control was a luxury, and every bit of it she had left was currently being used to keep body and flesh together.

"What? Of course. But why?"

"His depression was just getting worse and worse, and he really wanted to finish this project."

"You mean, the transfers?"

Christine nodded.

Seconds of heavy silence ticked by, one by one, only interrupted by occasional sniffles.

Finally, Christine said, "So, anyway. Here's the thing. I'm thinking, maybe I should do it too. Instead of months of brutal chemo and stuff."

More long, pregnant seconds.

"I can see how that could be a way out," Rachel said finally. "But doesn't that mean you'd have to *die* first?"

"Death isn't the hard part, honestly. *Dying* can be. But it can be over, quick and painless, you know? I can also be frozen. You know, cryogenic stuff. That's what Paul did, and…"

"Wait, Paul isn't dead?" Rachel seemed almost disappointed. In other circumstances, Christine might have laughed at her reaction.

"He's frozen, somewhere in Colorado." Christine finished the bottom half of her glass in a couple of big sips, letting that sink in. Then she looked at Rachel again. "Part of me is scared shitless, to be honest. It's never easy, is it?" After a few seconds of silence, lost in her own thoughts, she added, "I think if my mother could have transferred, it would have been so different."

"I guess." Rachel finished her wine, then frowned at the empty bottle. "For you, maybe. Not for her, right? She would still be dead."

"What other option did she have, though?" She paused and looked at Rachel. "Are you really dead when you're still there for others?"

"I…I never thought about it that way. I guess if you're dead you don't care anymore." Rachel took a gulp. "But what about kids? Robots can't reproduce, right?"

"I don't…I mean, no."

"But you've always wanted kids."

Christine's eyes welled up again. "Yeah, I know. But then with my cancer, there isn't much of a point trying to have a baby now, right? I guess that dream will have to die with me."

Rachel mechanically walked over to the kitchen with the empty bottle in her hand and grabbed a second one from the fridge. While she was away, Christine popped two opioids.

Rachel came back to the couch, unscrewed the metal cap, and refilled both glasses, then put the bottle in the ice bucket. She took Christine's hand and looked her dead in the eye. "Is Paul trying to pressure you into this?"

"No, I mean, not really."

"Stop saying that!" Rachel said, giving her a pained smile. "What do you mean?"

"He suggested it might make sense, and he can arrange it, but that's it."

"I'm not convinced, to tell you the truth." Rachel's voice deepened abruptly, as if she had just come to some conclusion. "Not at all. I don't think he'd be asking you to think about it if he hadn't done it himself."

Rachel had always tried to warn her about Paul. She had ignored the admonitions at her own peril, and she knew it.

"I guess that makes sense. But what options do I have, Rachel? Just wait to die? Wouldn't it be better to die *and* live?"

Rachel moved closer, took Christine's glass from her hand, and set it on the table, next to the massive Andy Warhol appropriation art book titled *Fair Use* they had picked up together at an exhibition at the Guggenheim two years ago.

"Christine, what do you really want, for you? Not for Paul, not for me. For you?"

"I…I wish I knew. I just don't want to…just die and… disappear." She started sobbing again, and then closed her eyes, her body almost limp from the wine and drugs.

Rachel barely managed to get Christine to the guest bedroom, where she fell into bed, fully clothed. Rachel lay next to her and put her arm around her, and within moments they were both asleep.

***

The next morning, Christine woke to the smell of coffee. After a brief detour to the bathroom, she followed her nose to the kitchen, where Rachel was putting the finishing touches on a beautiful breakfast spread.

"What are you doing?"

"Breakfast is served, madam." Rachel took her hand and pulled her over to the table, which held a huge, French-pressed pot of coffee, yogurt, freshly squeezed orange juice, a beautiful loaf of rye bread with Rachel's signature scoring of three Ss across the top, and five different types of jam.

"Oh, Rachel. This looks so wonderful!"

"Let's dig in!"

The coffee was perfect. She couldn't get coffee this good even at Eidyia with their super fancy Italian machine.

"What's this?" she asked, holding up a jar of jam with large chunks of ochre-colored fruit.

"Quince. I make it myself." Rachel beamed with pride. "Matt has managed to plant quince trees on his ranch. In a greenhouse."

"Quince? Never heard of it."

"I think it's a kind of plum. Big yellow fruit. Makes great jam."

Christine's eyes were sore and puffy, but she felt better, as if she had crossed some sort of threshold. She covered a slice of bread with fresh butter and added two big spoonfuls of the strange jam. The fruit was soft but resisted just a bit under the teeth, and the taste was like nothing she'd ever had before, yet it felt familiar. Plums, apples, apricots—what was it?

"This is amazing, Rachel."

"I know. When Matt brought me a bag of this fruit from last year, I tried to eat it raw and thought *yuck*. Then I read about how the Brits make jam with it and decided to try."

"I think I'm already addicted to it!"

"Say, on that topic, I was thinking: Can Transfers eat food?"

Christine swallowed and felt that harrying feeling of deep discomfort taking over her mind once again. "No. They can drink, but not eat."

"You mean, you won't be able to taste food? But you're such a foody."

"I know." The smile washed off her face. "I've been making lists of pros and cons in my head. Nonstop." She added another spoonful of jam to her bread and took a bite while Rachel refilled her coffee mug. "But when I think about it, it's really what I said last night. I have two options: Go through with treatment and probably die. Or avoid all that and live—not for me, but for others."

Rachel took a long, slow sip of the impossibly dark roast. She moved her gaze from the hot coffee to Christine's eyes. "When you think about living *for others*, what do you mean? What comes to your mind when you say that?"

"You sound like a therapist!"

Rachel smiled.

Christine put her bread down. She looked down and then up at Rachel again. "My mom. I think she would have wanted to live for us, for me and my dad. And I want to live for you, for my dad, and for her."

"And Paul?"

"Paul? He's a Transfer, so I'm not sure what living for him means, at least not anymore. For the Paul who's frozen in Colorado? Maybe. But that's a mirage. Who knows when he'll come out of there."

"Or even if."

"Right. So, all I can be is with someone, or some*thing* that behaves like Paul, and then that copy of Paul can be with a mirror image of me. It's really not about him or me, though. It's about everyone who stays."

"But we're all going to get old and die."

Christine's brain was still trying to make sense of it all, too. Yes, Transfers would be there for those that remained. For a while. But once everyone started transferring, Transfers would be there for each other. "I guess it's like we keep the memories of everyone alive, so collectively we all benefit. I'm hoping that Transfers will continue to grow, more or less like the person they replace."

"May our memories be blessings."

"What?"

"It's a Jewish saying. Sort of."

"I see." Christine took a sip of coffee. "I guess." She stopped trying to unjumble the mishmash of words and thoughts in her head. "As I see it, the option of avoiding all the bad medical stuff and being there for others just makes more sense."

"Looks like your mind is made up."

Christine pursed her lips and nodded. "Pretty much."

They sat there in silence, finishing breakfast, their minds traveling far afield in time and space. As they were cleaning up afterwards, Rachel put the gingham tea towel over her shoulder and took Christine's hand.

"Promise me you'll think this through some more before you decide, Christine."

"I will." She still hadn't finished crossing the gulf between knowing she could do it and actually doing it, so that was a promise she could make.

CHAPTER 26

The day before her flight home, Christine went to visit her dad in Huntsville. The house she grew up in was no longer immaculate. Her dad had always been so meticulous in maintaining it, but now paint that had been white in a previous life was chipping off the sideboards and peeling off the shutters, a victim of the caprices of time and the increasingly harsh Southeastern climate. Green mold had started to cannibalize the picket fence.

As she pushed open the gate, she saw him, an old man on his knees, holding a gardening fork, his back curved, haplessly trying to pull weeds, looking like an old toy about to run out of batteries. Her gaze moved a few feet away to the stone marking the small grave of her cat, Mowgli, crushed by a car the day after her eighth birthday. Her mom had gotten him for her at the Huntsville animal shelter when she was a toddler.

"Dad, it's me."

The old man turned around, and a smile more luminous than a thousand suns lit up his furrowed face. "My lapushka, it *is* you! Come give me a hug."

A tear went down his cheek, and as she hugged him, he felt fragile in her arms. The man she remembered as tall and athletic had become frail. Her dad could build rockets, and he had also built a cabin near Gulf Shores entirely made of wood by himself, using old-style mortise and tenon joints. Now she wondered if he could hit a single nail. Pangs of guilt gripped her throat for all the years she'd so rarely come to see him. She tried hard not to cry.

"Come inside, my lapushka."

As she entered the house, she turned a wistful eye toward the living room. Even with midday light pouring into the room, it could not bear a more inaccurate name. Everything was stale. The room felt unused and almost fake, like an exhibit in a historical museum: the old sofa redolent with decades of dust, frames of yellowing pictures, and curtains that hadn't seen their original color since the Beatles' first Liverpool concert. Then she saw the motorized chair attached to the wooden winder staircase.

He followed her gaze. "Ah, yes. I'm not getting any younger, you know."

She looked around at the dozens of mementos and pictures on the walls all around her, vestiges of decades of life in this old house. The picture of her state showjumping championship on Czar Alexander, the beautiful black Dutch warmblood gelding she'd trusted with her life more than once. Christine and her mom walking on the beach near Gulf Shores, holding hands, wearing huge sunhats and almost identical blue dresses with large dahlias on the back. That one was taken just a month before her mom's diagnosis. Her father receiving the Engineer of the Year award at Fischmann Aerospace, looking so proud. He'd been the best dad in the world that day. All this had once been part of her own life, but now it felt distant, foreign almost.

What will I do with all this when…?

"Unsweet tea?" he said, offering her a large glass with ice cubes and a slice of lemon

"Ah. Yes, thanks, Papa." Her mother had always insisted that she call them mama and papa, as Russians do. Her mother, steeped in Russian classics, had called her lapushka for as long as she could remember, and her dad had soon adopted the pet name as well. It made her feel like a daughter, all the stages of her life with her parents tied together by that word like a long string crisscrossing time.

They sat down in the old kitchen, on chairs with rusty metal legs and shiny green vinyl covers that had patches of discolored orange cushioning sticking out. She looked at the three-year-old calendar on the fridge, surrounded by too many handwritten notes to count. The musty smell of a kitchen left to slowly die made her yearn for lost time.

"Do you cook for yourself sometimes, Papa? Doesn't look like much happening in here."

"Not much these days, my lapushka. There is a service that brings me food once a week. I just keep it in the fridge and zap it when I'm hungry. But tell me, to what do I owe this surprise visit? When were you here last? Two years ago?"

She nodded and then told him about her work with Eidyia. He listened intently, his engineering background providing enough of a foundation for him to get the gist of it. He moved up on the old kitchen chair, eyes bright with interest.

"This is so exciting, my lapushka. I'm so glad you're building the future. I'm so proud of you."

She moved closer to him and took his hand, feeling the soft, dry, parchment skin, and trying as best she could to hold back the tears. Then she let out a breath and took a big sip of tea. "Papa, there is something else I need to tell you. I am dying of cancer."

The smile and all color disappeared from his face in an instant. In seconds, he looked ten years older, if that was even possible. "My lapushka. I don't... Your mother..."

"Papa, don't be sad. That's why I told you about Transfers first. I have decided to transfer." Saying it felt right. She couldn't let cancer do to her what it had done to her mom, and to herself, and her dad.

Her father stared at the table, immobile, a small tremor in his right hand.

Christine took his hand again. "Papa, I can be with you, *forever*. We can finally catch up. I miss spending time with you. We can get it all back, in a way. And Papa, you aren't getting any younger, as you yourself said. Maybe you should transfer too? I lost Mama. I mean, we both did. I don't want to lose you too."

He looked pensively at his glass of tea, which he finished in silence. Finally, he said, "Will you stay for dinner, my lapushka?" His eyes pleaded with all the energy left in his frail body.

"Yes of course, Papa. With pleasure."

His back straightened, as if invisible angels had lifted him up, in frame and spirit.

While her dad ordered food delivery, she went up to her old room. Nothing had changed here either. The dozens of ribbons and trophies from her showjumping days were all there. The picture of Czar Alexander with a young Christine standing right next to him in all her riding gear—those shiny black leather boots, a helmet way too big for her little head—holding her whip and prouder than any Olympic gold medalist. How she wished she could just go to the stables and talk to him. She could tell him anything. Anything. Looking in his eyes, she'd known he always understood. And that picture of her mom, serene, gracious, timeless. Christine felt her presence and was more convinced than ever.

Transferring was the best option, a way to keep those memories alive.

She would not let death erase it all.

CHAPTER 27

On June 15, 2038, at a ceremony surrounded by church and industry leaders, the President signed into law the People's Protection Act (PPA), making it more difficult for people to transfer. The Eidyia management team watched on a big screen in the usual meeting room at HQ.

As soon as the media event was over, Bart called their DC law firm. The lawyers suggested bringing a constitutional challenge under the Fourteenth Amendment: The Constitution did not define "human," and in *Smith v Alabama*, the 5–4 court decision that had expanded on *Dobbs*, the Court had recognized a fetus as a person with full human rights even if that fetus was not able to survive outside its mother's body. It had ruled that a human being existed as soon as the transition from embryo to fetus was complete. Critics had been quick to point to the stupidity of that test as a biological matter, but the Evangelicals' big dream had finally come true. Now it might come back to bite them. The test meant that there was no need for a person to be fully formed or even viable.

The team also conferred with the lobbyists, who had spoken to many members of Congress, in particular the eighteen members who had applied for transfers. From their perspective, pressure from the hospital and pharmaceutical conglomerates was simply too great to fight back. They urged Eidyia to release a new code of ethics at a big media event.

After a long discussion, it was decided that a court case would not be a good option, at least not at this stage, and the group ultimately agreed to have a major event in DC in a less than a month to launch their Code of Ethics for Transfers. The Court of public opinion was the right forum.

***

Eidyia booked the ballroom for the evening of June 30 at one of the best hotels in DC and invited business leaders, journalists, civil society, all members of Congress, and many others. Thousands of members of the public had applied to attend, and one hundred had been selected at random. The room was packed, and dozens of flashes went off every second.

Bart and Paul stepped onto the stage. Bart had always resisted the tech CEO uniform at big public events—sneakers, tight jeans, t-shirt. Instead, he sported tailored wool pants, a black turtleneck, and a jacket. Paul didn't buy into it; Valley normcore would do here as it did everywhere else.

"Codes of ethics have a long history," Bart began. "Some, like the Hippocratic Oath, have almost sacrosanct status. There have been many codes of ethics for AI and robots, including Asimov's famous Laws of Robotics. His first law was a directive to robots to 'do no harm,' but we know that robots can cause harm—sometimes to prevent greater harm, sometimes by accident. We have moved well past that. We have taken a giant leap. We are now able to create beings that will be, if not human, then just like humans. Beings that look like us, think like us, in fact do almost everything exactly like us. We can transfer a person's thoughts, their *mind* into this new being. This new population that does not depend on increased food production and that does not get sick is a huge step forward for all of us."

He paused, to muted applause.

"This way, we can live forever."

Another pause, but this time a slow wave of applause started, and within thirty seconds, about two thirds of the room was roaring and on their feet, but not the professional journalists in the first few rows. Bart signaled for everyone to sit down.

"This requires some rethinking of our ethical practices. Today, we are here to unveil the rules that Eidyia will apply to Transfers. Let me turn it over to my friend and colleague, Paul Gantt."

Paul moved to the center of the stage, tablet in hand. As he read, the rules appeared in giant letters on the screen behind him.

Rule #1: A transfer is only possible at the time of, or promptly after, death. Each person can only request one transfer, and only for themself.
Rule #2: Only living persons of sound mind can request transfers.
Rule #3: Eidyia will price each transfer based on the requesting person's ability to pay.
Rule #4: A person may request physical changes as part of the transfer process, but only for functional reasons. No other changes are possible.

A hundred hands went up. The Eidyia team had received an attendance chart from the lobbyists, and Bart consulted it before pointing to a journalist in the front holding a huge tablet and wearing a purple jacket and large glasses.

"Yes, Stapleton-san, here in the first row."

"Are you saying that you will not bring back people who are already dead?"

"That is our position. We could go back a few years at the most, in any event, given that the technology necessary to capture enough data is relatively new."

"So, if I may follow up, you're forcing everyone to wear an S-Chip?"

"We are not forcing anyone to do anything. Transfers, like the S-Chip, are entirely voluntary." Bart moved to the other side of the stage. He pointed at an elderly and very well-known *New York Times* journalist in the third row. "Cooper-san."

"What is your production capacity? Can you accept all transfer requests?"

"We are developing an algorithm to prioritize applications." Bart was trying to look relaxed, but Paul knew him well enough to notice that the pitch of his voice was higher than usual.

Cooper was not done. "An algorithm? Based on what kind of criteria?"

"We've actually asked our AI engine to suggest appropriate criteria."

"Will you make those public?"

"We have not decided that yet, but I can tell you that the criteria are based on an evaluation of an individual's contributions."

Cooper wouldn't let it go. "Can you give us an example?"

"Yes. A person convicted of murder, rape, spousal or child abuse, or other major crimes would have a very low priority."

"So, you're basically *rating* every American? Do you think the American people will accept that?"

Bart was still looking at Cooper, forgetting or ignoring the PR team's advice not to engage in a conversation with any one member of the audience. "First of all, companies rate people all the time. Think of insurance companies or credit agencies. Second, we are offering a service, and we cannot offer it to everyone who might want it, at least not for the near future. Having an algorithm with clear defensible criteria is better than picking at random. Remember that financial capacity will not play a role in our decision to offer the service to anyone."

Paul watched the sweat run down Bart's temples. He was probably running out of steam, and Cooper had had the floor long enough. Paul looked at the attendance chart and pointed to the second row. "Wilson-san?"

Janet Wilson was only twenty-eight, but she was a fast-rising star at CNN who covered all things AI. "Can someone be, well, I'm not sure how to put it exactly, but can a person be stored until they make a decision?"

Paul nodded. "Yes, in a way. Transfers will not happen exactly at the moment of death. We will wait for death to be confirmed, which means there must be some form of storage of the form– the transferor's personality."

"Will you store everyone by default?"

"Technically, we can do that for anyone who has been using an S-Chip for long enough to generate the amount of data we need."

Suddenly, someone in the back yelled, "Do you think you're like some sort of gods? You decide who lives and who dies? You will all go to hell for this. Your algorithm is like some sort of Last Judgment but from Satan. It's God's job to decide who gets eternal life."

Many people in the back of the room professed their agreement, and some got up from their chairs.

Paul stared at them a moment. "We are definitely not gods. We are a technology company. We certainly do not pick who lives or dies. People die. That is inevitable. When they do, they now have an option for their personality to continue to exist." Paul sensed a wave of commotion going through the crowd, as if a safety valve had just been tripped. The room was turning against them. No point trying to convince these dumb humans. "Thank you, everyone."

With that, he tapped Bart on the back and they exited the stage rapidly, practically running for the exit. A heavy metal door clanged shut behind them.

CHAPTER 28

A week later, Bart, Jeremy, Koharu, and Paul met at Eidyia headquarters early in the morning to review the results of Iwa, the Eidyia algorithm programmed to rank applicants for transfers. The algorithm had identified two pillars to base its decision on: human development, including a reduction of human suffering, and the amelioration of the environment. The algorithms had parsed data from 400 million S-Chips and identified those individuals who scored highest based on their contribution to the two pillars. Based on the findings, the computer had graded the subjects. Only 36 million made the highest cut: A. Another 52 million were ranked as A-, 61 million got Ds, and almost the same number Fs. The rest were a mix of Bs and Cs.

"This is interesting," Paul said, putting his tablet down and smiling at the others. "It picked people who will develop as individuals in ways that make them make positive contributors to the collective good. Truly excellent."

Bart picked up his coffee mug. He was already on his fifth macchiato. "Well, that's not surprising."

He was right. The idea that letting people pull the covers their way as hard as they can and everyone will be better off as a result had produced a lot of wealth, but even more inequality, and it almost irreversibly destroyed the planet. Capitalism had always assumed infinite resources, but obviously resources are not infinite. Any rational analysis leads to only one conclusion: this is economic nihilism, a system that cannot support both growth and a dying ecosystem. Problem is, that ecosystem is everyone's essential support system.

But when technology starts to think on its own, the game changes. Radically. Iwa was proof of that. Humans can be irrational. It is deeply idiotic to use a finite ecosystem as if it were infinite. Iwa could see how things had begun to change and probably wondered why humans got the planet to this breaking point when all the signs were there for decades.

Jeremy got up from his chair and walked to the coffee machine. "What I know is that we can't process all 36 million. There's an order of magnitude problem as of right now. We promised 28,000 R-S units to the DoD by the end of next year. That's a more realistic scale."

179

"Well, we'll have to see how many of those who applied are in the 36 million," Bart interjected.

Jeremy moved back to the table, picked up his tablet, and tapped on the screen. "That's easy. With the most recent data, 536,780. That's still *way* too many for now, and that was just a test with 400 million. If we run it with *all* S-Chip wearers and give people more time to sign up, we'll get millions of requests to process."

"Well, they do have to die first," Paul pointed out.

"True, but we're still looking at a huge ramp up problem."

Bart was staring down at his tablet, not following the conversation. His face suddenly sank. "I'm afraid I have some bad news. We just got word from Jim Goodman that another bill has been introduced in Congress to make transfers entirely illegal, including existing ones. Fortunately, it seems that the bill drafters are struggling to define what exactly the illegal operation is, which should give us time to react. I think our friends there may be the ones helping, but if it comes out that they have asked for transfers, I bet they'll be excluded from the discussions somehow." Bart's face had turned red, and a big vein was visible on his right temple. He set down his cup with a bang.

Paul was unphased as he scanned Goodman's message. "I'll call General Armstrong and see what she makes of it."

The others were all staring at Bart. He so rarely lost his nerve. No one spoke.

Finally, Bart took a deep breath. "Let's reconvene on Friday afternoon."

As they dispersed, only Koharu stayed behind, staring into their cup as if looking for answers in their tea.

CHAPTER 29

On Thursday, July 8, 2038, Christine landed around noon in Denver, where Paul was waiting for her. She was pale and anxious and her hands wouldn't stop shaking.

"Are you sure you're ready for this?" Paul asked as they walked to the PC garage.

"Scared as hell, but ready."

They drove to the Flatiron Cryonics facility about twenty miles outside of Denver on the road to Boulder. On the way there, they were both silent. Christine looked out the window at the snowcapped mountains on the horizon. She had hiked there with Rachel on a trip to Boulder in college. Royal Arch, the trail was called, and it had felt like going up the Eiffel Tower. The view from the top was magnificent. It had left an indelible imprint in her mind, in the small catalog of memories that remain when all else is forgotten.

When they arrived in front of the glassy modern building, they were met by a robot who showed them into a dimly lit room with quiet spa music playing faintly in the background. Four chairs clustered around a glass table that held a stack of cups and a pitcher of water.

A tall, skinny woman wearing a white coat and holding a tablet walked in. "Christine-san, welcome."

Christine looked at her name tag: Chantal Stack, M.D.

The woman sat down across from her. "I am Dr. Stack. Please, call me Chantal."

Christine's lips drew a faint smile.

Dr. Stack smiled back. "I will oversee the whole process today. I see you already signed the forms online. Do you have any questions for me before we get started?"

Feels like the dentist's office. Geez. Christine picked up the pitcher on the glass table and poured herself a cup. "Can I see what it looks like...I mean, where do you keep the bodies?"

"Of course. Please follow me. Paul-san, will you be joining us?"

Paul nodded and got up and they followed Dr. Stack through a maze of immaculate corridors, where they encountered a few people, all wearing white coats and smiling too much. They eventually arrived in front of a door marked *Observation Deck*. Dr. Stack tapped a card on a small black plate on the wall, and the door opened.

"After you."

Dr. Stack followed them in and pushed a button on the wall. The metal screens covering the windows started to lift. What Christine saw astounded her. Thousands of shiny gray pods, made of some sort of metal, covered an area the size of a football field. Christine could see three or four small robots whizzing around quickly.

"We have three floors like this," Dr Stack said.

Christine looked at her, then at Paul. "Where is the human Paul? Can I see his... pod?"

"Of course." Dr. Stack tapped on a screen on the wall and a robot rolled toward the back of the room. "Watch the screen."

A picture appeared, and Christine read: *Gantt, PF-5272-1*. The robot moved away from the pod and went all around the metal chrysalis with its camera. Ice had formed on a small window where Christine imagined his head would be. In her mind, she put her fingers on the window and the crystals started to melt until Paul's softly smiling face appeared. A dam began to crack inside her. This was all too much. She started to cry.

Dr. Stack tapped on her watch. A robot entered the room carrying a glass of water and offered it to Christine. She reached for the glass slowly and took a sip. A minute later, she started to feel lightheaded. Then the lights went out.

CHAPTER 30

The next morning, Christine Prime joined the Five to talk through the implications of the latest bill as planned. It was going to make all transfers illegal, with one exception. General Armstrong had ensured that the definition of "Transfer" in the new bill would exclude the R-S. She had clinched the deal with key senators when she'd said that, according to some intelligence reports, China was developing a competing technology. That was the only bit of good news. The bad news was that Big Pharma, several church groups, and the hospital lobby, whom all stood to lose billions over the coming decades, had created a 100-million-dollar fund, labeled the Humanity Initiative, to fight Eidyia.

Bart banged his fist on the table. "Stupid fools!" he shouted. "Don't they see how great this technology is?"

"I am threatened by them; I want humans to win," Christine recited.

"What?" Bart asked.

Christine turned her chair towards him. "It's a line in a Tarkovsky movie. A Russian film director. No matter. The point is, I'm not surprised. Plenty of sci-fi novels and movies have suggested that humans will always be scared of anything that threatens their dominion over nature and other species. It will take years of successful coexistence for that fear to abate."

"But people, *We* the People, *want* transfers," Bart said.

Christine kept her eyes on Bart. "Yes, their desire for some sort of immortality may well trump their fear of coexistence. Besides, transfers happen at the time of death, so what do they care? *Après moi, le déluge.*"

A huge smile suddenly appeared on Bart's face. "Deluge. That gives me an idea. Let's go for a sea change."

Everyone looked at him, mystified.

"I was just trying to be clever. Sorry, guys." He was still smiling like a little rascal too proud of his find. "I meant, let's go offshore."

"Offshore? Like Liechtenstein?" Jeremy asked, frowning.

"No, I mean *literally* offshore, as in off-the-shore. Let's build a fucking island."

Silence.

Paul looked at Christine. Then he turned towards Bart. "I think I like it. But where?"

"In international waters, of course. Then we could do what we want, right?" Bart turned to Christine. "What would that mean legally?"

"I'm really no expert in that field, but as far as I know, other than basic international law, you could do pretty much what you want there."

Bart got up and paced the room. "So, Congress couldn't stop us, right?"

"They could pass laws to regulate what Transfers can and cannot do in the United States." Christine's voice remained unchanged. "They could also ban their entry into the country, I suppose."

Jeremy looked at her curiously. "Yes, computers might tell the border service that the person entering the country has died."

"If the government knows," Paul said.

Jeremy looked at him enigmatically, clearly unconvinced.

"I say we do it," Bart said. "Process transfers, and then let the Transfers and people who want to transfer argue their cases, in courts or Congress." He was as excited as an eight-year-old who just got his first real bicycle. "I will get a feasibility study started right away, and a budget."

***

Two weeks later, the Five sat around the campfire in the backyard of Paul's Jackson Hole mansion enjoying a few bottles of Barbaresco. The air was crisp and the moon cast its milky light over the large grassy meadow behind them. The wind made the tall grasses dance. Bart, Koharu, and Jeremy watched the hypnotizing waves, but Paul seemed distracted, and Christine sat in silence, watching him.

Bart was the first to speak. "It seems the more we think about the implications of Transfers, the more stuff comes out of the woodwork."

Christine raised an eyebrow. "Not really surprising, is it? It takes the law years to solve any question that deals with death. Think of the endless debates about the right to die with dignity."

"That was paternalistic junk, just like this bill about Transfers," Bart replied.

Jeremy put his glass down. "Well, arguably any law is paternalistic because all laws are meant to interfere with liberty."

Bart nodded. "I see your point, Jeremy. I guess anything that interferes with my ability to decide about what happens to me is unjustified, isn't it?"

"That argument usually resonates very loudly, especially in America." Christine took a small sip. "I'm actually worried about how the legal system will respond otherwise."

"What do you mean?" Jeremy asked.

"I mean, will the law consider Transfers like persons? Will they be able to function like humans, have bank accounts, be liable if they do something wrong, get married, and so on?"

Bart frowned. "Now I'm afraid Congress will regulate that, too, unless they make all of it illegal of course."

"It's not that clear that they can." Christine swirled her wine pensively. "First, states make most of the decisions about what makes a person a person, not the federal government. Second, if courts consider Transfers as people, they will have constitutional rights that limit what Congress can do."

"So how would that work in this case?" Koharu asked.

"I could see the Equal Protection Clause kicking in," Christine said." If you agree that Transfers are persons, or that they should be treated as such, then this might apply to them. Saying that they could not get a job or open a bank account, for example, would be illegal."

"I think we're going to make some lawyers very rich," Bart quipped.

"Maybe one day those lawyers will be Transfers," Paul said, giving Christine a sardonic smile.

***

The next morning, as they convened in the vault, Bart explained how, for $19.5 billion, they could transfer the entire Eidyia headquarters to a newly built island. Given that Eidyia had more than that amount in cash, plus the significant tax savings of moving the company offshore, Bart recommended leaving the United States and moving forward with the idea.

"And how long to build it?" Jeremy asked.

"Looks like it will take eleven months to get the island off the ground, I mean off the seabed, and then another fourteen to sixteen months to get everything ready. If we work twenty-four-seven and throw tons of equipment and money at it."

"So, we don't ramp up for, what, two years or more?" Jeremy asked, visibly upset.

"Luckily, we have another option," Bart continued. "The Office of the Prime Minister of Canada reached out to our lobbyist friends last week. They would be happy to have us in Canada. They can even provide inexpensive power if we move to Quebec." He flashed a smile at Christine. "You'll be able to practice your French!"

She smiled back, faintly. This came as no surprise. Canada had been a leader in AI for well over twenty-five years, and the discussions on a Third Montreal

Protocol had just started. Some proposals on the discussion table included rules for technology like Transfers.

"And how long to get started *there*?" Jeremy asked.

Bart tapped on his tablet. "They have a number of empty refurbished warehouses in Montreal, apparently. We just need to install the equipment. I think we could be up and running in four or five months. In the meantime, we will continue to produce the R-S units here since this won't be prohibited."

"Won't we face the same political difficulties in Canada?" Koharu asked.

"It doesn't look like it," Bart answered. "They have a parliamentary system, and the prime minister's party can block any legislation they don't like. Religious groups also have much less political influence there."

"It's like the debate over stem-cell research," Koharu said.

Christine nodded. "Yes, exactly. Canada is much more interested in AI than in pharmaceuticals. Their hospitals are operated by the government and paid for by tax dollars, so in their case having less to spend on healthcare is a plus, not a minus like here."

After a lengthy discussion, they agreed to make a deal with Canada and start moving the equipment to Montreal as quickly as possible, but also to start building the offshore island, a project they named Argo. Neither project was to be made public now, although the move to Montreal would be hard to keep secret for long.

CHAPTER 31

Ironically, the InterContinental hotel Paul chose for their first visit to Montreal that summer to keep up appearances still had human check-in agents. Their room on the twenty-third floor had a view over downtown and the "mountain"–as Montrealers like to call the *Mont-Royal*, the hill in the middle of the city, and after dropping off their stuff, Christine and Paul immediately went out for a walk. The hotel was within a stone's throw of Old Montreal, a part of the city originally built in the seventeenth century when Quebec was called New France, but partly rebuilt to British architectural standards after it became a British colony in 1763. In front of them stood the Notre-Dame Basilica with its dark blue windows illuminated. They walked down a cobblestone street and arrived at a large park on the St. Lawrence River.

Paul pointed to a stand selling something called beaver tails. "What are those? Sounds gross. And very chewy."

She gave him a strange look, knowing he could pull the information up on the Grid just as easily as ask. "Beaver tails? It's not actually from a beaver. They're made of fried dough, topped with lemon, maple syrup, chocolate, and other goodies. They're not bad, actually."

They continued walking, and a few minutes later, arrived in the Place Jacques-Cartier, a square full of buskers and artists selling caricatures and paintings of Old Montreal.

"Where are we going?" Christine asked.

"I want to show you the building we're moving to. It's within walking distance."

They headed back to the waterfront and then east to a newly built area. Paul pointed to a vast red brick building that looked like a huge warehouse, with a tower on top, a smaller tower on top of that one, and a gigantic analog clock on that.

"Is this it?"

"Yes, we have half of it for now, but the building can be expanded towards the river if we need more space. It's a convenient location. There's a train to the airport and access to the subway close by."

"This is obviously not a new building."

"No. It used to be a brewery. First brewery in Canada or something like it. There's still a very faint smell of beer inside."

"*Beurk*," Christine said, making a funny face. "Christine never liked beer."

Paul turned to her abruptly. "What did you say?"

"*Beurk*. It's French for 'yuck.'"

"No, the other thing. You know you should *never* refer to Christine in third person. Even if it's just with me."

Christine flinched. "I'm just still not used to the idea that I have to be her. Even less so that I *am* her."

They continued walking in silence, passing a series of eighteenth-century buildings made of gray stones. Christine mused about how humans keep vestiges of their past, as if visible signs of human history made their lives fuller and deeper. It was just an illusion, of course. Old cultures resting on their brick, stone, and mortar laurels eventually faded like the pages of an old book. But the attachment to tradition was strong, a feeble sentinel to protect against the ephemerality of human life.

They turned on rue Saint-Denis and ended up in front a chichi wine bar in the Plateau, one of the city's older but trendy areas.

"Let's go in, for old time's sake," Christine said.

Paul looked at her, shrugged, and then turned to open the door.

A human server showed them to a table and handed them menus. "Do you know what you want, or should I come back?"

"You pick, Paul."

He scanned the list quickly and picked a 2029 White Hermitage. When he looked up, a flash of concern crossed his features. "You look sad, Christine."

He was right. The forlorn thoughts of never again relishing noix de St-Jacques aux têtes de violons or wild Atlantic salmon served with *confiture aigre-douce d'atocas* was showing on her face. "I so wish I could enjoy the food. You remember in Lecce, Paul and Christine ate this divine dish, *orecchiette alle cime di rapa*?"

"*Not* Paul and Christine," he hissed. "You must think of it as *we*. And yes, I do remember. With a bottle of Susumaniello."

The server brought the wine, opened the bottle, and poured two glasses.

When they were alone again, Paul continued, "I was like that at the beginning too. It took a few months to pass. Then the many advantages of being a Transfer started to become clearer. The biggest one by far is when you realize that humans waste all this time trying to *be* when all we need to do is exist. It's so much better. Mind you, I still enjoy coming to places like this for the atmosphere. Congeniality, I guess." His eyes flitted back to the menu. "Say, what are these *têtes de violon*?"

Paul used his connection to the Grid to pull up a translation. "I know the literal translation is fiddleheads, but that obviously cannot be right."

"The Grid isn't all that perfect, it seems. They're fern shoots." Christine took a sip of wine and looked around.

The bar was an eclectic mix. It had a high ceiling embossed with tin that probably belonged to the original building. Late nineteenth century, she guessed. The room incorporated a stage for live music, which held only a drum set, a yucca, and a snake plant, both of the latter ensconced in brown bamboo pots. Behind the bar were hundreds of bottles of alcohol of every size and color standing on backlit, blueish glass shelves. The kitchen was open, and an army of cooks wearing chef's hats ran around like well-organized ants making order out of apparent chaos. The tables were made of reclaimed wood, according to a sign on the table itself. Lounge music played in the background.

She turned her gaze back to Paul. "What do you mean when you say advantages?"

He smiled. "Well, you just mentioned the Grid. That's a biggie. Like being online without having to use any tool. But there are many more. You don't have to go to the gym anymore."

That resonated. Gyms had always been a love-hate thing for Christine. Everyone judging everyone else. Comparisons on steroids, pun intended. Something you do for the after-gym feeling.

"One of my favorites is silence," Paul said. "Think about how we walked for hours today and sometimes didn't exchange a word for half an hour or more."

Christine nodded.

"It's easy for us. Natural I would say. Silence terrifies humans. It's just a void that needs filling. But we know we're connected, with or without words."

Christine was silent for a moment, savoring his words. Then she said, "I must say, I feel frustrated sometimes when I see how humans make decisions."

"It won't get any better, I can tell you that." Paul smiled, but there was a tinge of vindictiveness in his eyes. "We *know* how they think because we can think like them. And with their brain implants they think they can think like us. But they can't. They have innate limits, and we don't. Or not the same. And it's a good thing."

"What do you mean?"

There was more than a twinge of annoyance in his voice when he said, "Sometimes I see humans like a fragile patchwork of neuroses barely held together by a glue of cognitive biases. You know how it is. They can't make decisions to save themselves even. A single leader can undo decades of progress."

Christine pursed her lips. That struck her as correct, but still stung. "I am beginning to see that." She fell silent as the server came by their table.

"*Autre chose?*" he asked, pointing at the empty bottle.

She smiled at him. "Non. *L'addition, s'il-vous-plait.*"

After settling the bill, they exited the restaurant and walked back down rue Saint-Denis all the way to the Old Town. It was a refreshingly cool evening, and as they walked, Christine tried to hold hands with Paul, as they had often done before, but he found every excuse to pull his hand away. Did his desire to keep his hands empty match his heart? Christine was adrift in waves of strange new feelings. Not disappointed as such but frustrated that behavior that had felt automatic and natural was no longer reciprocated.

Paul interrupted her thoughts. "You know, I did take French in high school, and even a course in college, but I'm having real trouble understanding people here."

"They do speak a different kind of French," Christine said, putting on her professor hat. "Some of it is called joual. It probably comes from the word for horse, cheval. For many years, people in Quebec only had access to English-language instruction manuals for everything from their cars to their vacuum cleaners. And at the time, most employers were English-speaking and used English words for everything in the workplace. So, the French-speaking population made up words that were based on those English words, but they turned them into something more French sounding. Meanwhile, people in France are using as many English words as they can, especially in tech and business."

"The French think everyone else is stupid." Paul's tone made his disdain clear. "They used to dominate everything, but they haven't realized that times have changed. All they still have from the days of their empire is their arrogance."

"You don't like them. Just like Paul." The memory of Paris brought a smile to her face.

Paul gave her an exasperated look. "I *am* Paul. And when I was in Paris, people made fun of my bad French, but it was usually better than their nonexistent English!"

"They're not all like that!" Christine protested. "Don't judge the whole country because of a few stupid people. There are plenty of dumb Americans too."

"True. Stupidity is one of those human virtues that was spread very evenly. I get that more than ever now, and you will too."

They walked two more blocks in silence, crossed a street called Roy, and passed a large, gray stone building on the left. One of the few buildings on the street that was neither a stylish restaurant nor a trendy shop.

"There's something else I cannot quite figure out," Christine said after a while.

"What is it?"

"Well, when you live forever, you gain more knowledge than humans. Any human. A fifty-year-old may think that twenty-somethings are ignoramuses. Imagine if instead of being forty or fifty, you were a hundred and fifty, or two hundred and fifty. Wouldn't you consider fifty-year-olds or even eighty-year-olds as poorly educated as children, and treat them as such? Wouldn't you naturally be the one with all the wealth and power at that point? Does that suggest putting some sort of expiration date on the Transfers? I mean, otherwise, they—" she lowered her voice, "*we*, will become overlords. It's written in the stars."

Paul stopped walking, turned to Christine, and grabbed both her hands. "Look, maybe that would give us a sense of mortality and maybe make us more human, but it defeats the purpose."

Christine tilted her head slightly. "What purpose?"

"You'll begin to understand. Soon."

He started walking again, she followed, and they soon arrived in front of a small park. *Square Saint-Louis*, the plaque read.

"Square?" Paul said, looking at Christine. "Isn't that a *carré* in French?"

"Mmmm. I guess they're trying to sound like France," she said. "Use as many English words as you can but pronounce them the French way."

A minute later, as they were passing in front a Greek restaurant packed with people, Paul said, "I can see why humans like this city. I hope it will work for us. For now. And that your meeting with the dean goes well tomorrow."

"I hope so too," Christine said, a vague air of sadness shrouding her eyes.

They walked to the middle of the park near a big fountain and sat on a bench. Nearby, an old man wearing a woolen hat that had seen better days, baggy pants, and a flower-covered shirt that would have fit a clown in a circus act was pushing an old shopping cart, surrounded by a group of pigeons like a drunk infantry division following a tank. Occasionally, the man threw a handful of birdseed at the birds.

After watching him for a few minutes, Christine asked, "What do you think is scarier, death or knowing that you'll live forever?"

"Nothing scary about living forever. Death is so…how can I put it? Human. That is one good reason I'm not sure I want to be human. Let me rephrase that: I am sure I don't want to." He got up. "Let's keep walking."

They quickly found themselves on a pedestrian street full of restaurants.

"This will affect the Transfers' entire life philosophy." Christine said after a while, looking distractedly at the crowd around them, then at Paul. "There's a line in *Solaris* where Rheya says, 'I don't believe we are predetermined to live our past. We can choose to live differently.' I guess this is what's bound to happen, right?"

Paul smiled enigmatically. "You're beginning to get it."

***

When Christine returned from her meeting at McGill University the next morning, Paul was reading something on his laptop. He looked up as she came through the door.

"How did it go?"

"Not sure, frankly. The dean wasn't overly enthusiastic."

He turned the screen to her. "In other news, the story is out."

She glanced at the headline as she took off her shoes. "Who leaked it?"

"No idea. Probably someone in the Prime Minister's Office trying to score PR points. I've already received more than a dozen interview requests. It's a good coincidence that we're here because I'll be able to do them in person."

Christine's watch buzzed. The McGill dean, asking her to call as soon as possible. She showed Paul, then dropped into the chair next to him to make the call.

The dean picked up almost immediately. "Professor Jacobs? Thank you for calling me back. I just saw the news. About Eidyia moving to Montreal, and your name is mentioned as a special adviser, is that correct?"

"Yes. I've been working with them for a while, on legal and ethical issues."

"That sounds extremely interesting. Well, I have some good news. I've spoken to the central administration, and we would be happy to have you here at the faculty of law to teach AI Ethics. We will match your current salary and throw in fifty thousand per year as research funding. How does that sound?"

"I'm...I'm happy to accept," Christine said, struck by the gulf between the cool reception in the dean's office earlier and the offer she had just received.

"Excellent. Will you be able to start in the fall?"

"Oh, that's quick. But I think so."

"Great. I'll have HR send over some paperwork."

They said their goodbyes, and Christine ended the call. She stared at the phone, dumbstruck, then turned to Paul. "Wow. I guess she was just keeping her cards close to the chest."

Paul looked up from the screen. "Maybe, maybe not. She is a dean. You know."

"Ah, yes. Of course. She thinks Eidyia will make a big gift to McGill!"

"Bingo! And they *are* lucky to have you. A double win for them."

CHAPTER 32

On her first day at McGill, Christine walked from the furnished apartment they had rented in the Old Town up to the law school, a twenty-five-minute stroll through the downtown core. She had brought Harry and Dewey, but the cat kept hiding under the bed unless it heard Harry opening a food can. Christine felt strange at first not having Dewey to pet, but then nor was she watching Eikasa anymore. She attributed the cat's change of hearts to the move to her new digs.

As she was walking towards the Law School, she noticed people looking at her, usually for just a fraction of a second, and remembered how Christine used to adapt her behavior in response to what she thought people might mean by fleeting eye contact. Hell is other people. How silly that complete strangers can dictate your behavior this way, right? Yet that mirror through which we constantly see ourselves is undeniably there. Christine Prime was not in human hell but into, well, she wasn't sure exactly what just yet.

When she walked into the classroom, the collected students filled the room with a strange reverberation, a mixture of respect, admiration, and perhaps a fear of the unknown. They had read that Christine was the *éminence grise* at Eidyia. They also knew that despite her move—and a large part of Eidyia's—to Canada, she was still part of the US delegation to the conference on the Third Montreal Protocol on Ethics in AI. Not only was she a star, but US law professors had a reputation for harshness in their application of the Socratic method.

"I'm happy to be here with you today," Christine began. She repeated it in French. "*Je suis contente d'être ici avec vous.*"

With just those few words of French, the vibe in the room became palpably smoother. She was adapting to them, even of most the students spoke French as a second or third language.

"Let us begin by defining our subject. AI and robots. What are they, and what are they not? Do we define them by their own characteristics, or in comparison with something else, like, say, humans or animals? You will all be my teachers here. I am not well versed in Canadian law, so please help me learn."

That last sentence removed another layer of tension from the room. As a Transfer, Christine had instantaneous access via the Grid to multiple knowledge bases, but she knew asking would make the students feel empowered. More than that, at heart, so to speak, Christine was still an educator, and the mark of any good teacher was one who wants to listen and learn.

"How about we start with defining what a person is in Canadian law?"

A feminine person with short blond hair with blue highlights in the front row raised a hand.

"Yes, and please tell me your name."

"My name is Julie Dubois, she/her."

"Yes Julie-san, go ahead." Another silent wave rippled across the room. Canadians had begun to embrace -san before Americans and perhaps the students were not expecting it from Christine.

"The first thing you may not know, Professor, is that the Supreme Court, I mean the Supreme Court of Canada, of course, once said that women were not persons."

"You're kidding? When was that?"

"1928, I think. The Court had to decide if women could be appointed to the Senate—"

"Appointed? Aren't Senators elected?"

The class laughed, but not derisively, more nervously, to release tension. Christine tapped into the Grid and realized her mistake, but she let them explain.

"No, not here," Julie's neighbor, a young masculine person sporting a goatee, said.

"And what is your name?"

"Bindu Mathur, he/him."

"Yes, Bindu-san. So how do you become a senator in Canada?" Though she knew the answer, she thought playing dumb was getting her brownie points.

"You're appointed by the Prime Minister," Bindu answered.

"Oh, and for, what, six years?"

More chuckles.

"No, for life, I mean, until you're seventy-five."

"I see. Life until you're seventy-five. Okay. But I take it that women are now allowed to be appointed?"

"Of course!" Julie smiled. "The Charter—that's what we call our Bill of Rights equivalent—was adopted in 1982 and it guarantees equality."

"Ah, now this is getting interesting! Equality of what? Or maybe I should say between what?"

A hand went up in the back of the room.

"Yes, what is your name?"

"Bill Townsend, he/him. I go by Billy."

"Billy-san, do I detect a familiar accent?"

"Could be, Professor. I'm originally from Knoxville, Tennessee."

"Been there. You'll have to tell me some day how you made it all the way to McGill, but let's keep that for later. Do tell me one thing, did you study law in the US?"

"Yes. My major at UTK was pre-law."

"I see. Okay, and what can you tell me about equality rights in Canada?"

"Sure. So, the right of equality in Canada is not so different, at least on paper, from the Equal Protection Clause in the US." Billy looked down at his screen. "Article Fifteen of the Charter says, 'Every individual is equal before and under the law and has the right to the equal protection and equal benefit of the law without discrimination.'"

"I see. Thanks, Billy-san."

Billy beamed with pride, a smile that gets you to heaven without confession.

Christine turned to the rest of the class. "What is an individual? Why would an artificial being not be a person?"

"You mean like Transfers?" Bindu asked.

Christine took two steps towards him. After all the press reports, she had fully expected questions about Eidyia, so she was not surprised. "For example."

"We do not yet have a court case on that point, of course," Bindu said.

Julie raised her hand, and Christine turned and nodded to her.

"There was a case on fetuses, *Tremblay v Daigle*. It's an old Supreme Court case, from 1989."

"And what did the court decide?"

"It was complicated because the Civil Code here in Quebec gives fetuses some rights, but the rights can only be exercised if the fetus is born alive and viable. Then Quebec has its own charter of rights, and it gives rights to 'human beings.' In the end, the Court decided that a fetus was not a human being. The law is the same in Ontario, and in England, I think. Rights begin at birth."

Christine was struck at once by the maturity of the students, their willingness to engage so quickly, and the different assumptions about the law and the nature of human society on which it is erected in Canada and the US. She was happy to be there. No… She felt that Christine would have been happy had she been there herself. With a start, she realized she'd been staring at the ceiling for too long. Fortunately, law students often intercept that as a sign that the instructor is in deep thinking mode. She brought her thoughts back to the fetus issue.

"Yes, I was at a conference in London a few years ago and there was a case, *Paton*, I think, where a man was trying to get an injunction against his wife who wanted an abortion, and it was refused." She looked behind her. There was a gray desk made of some unidentifiable synthetic resin attached to a small podium with a tablet. That would have to do. She walked to the desk and sat on the corner. It was higher than her usual classroom desk, and she felt a bit off-balance. But this was Christine's favored position, so she decided to stick with it. She scanned the group from right to left. "What about corporations? Do they have rights in Canada, I mean as persons?"

Julie raised her hand again. *Eager student that one.* "Go ahead Julie-san."

"The terms 'everyone,' 'citizen,' and 'human being' are used in the Charter. For example, it uses 'everyone' for free expression rights. The Supreme Court said the term can include natural and legal persons."

Christine took a step towards Billy. "So, Billy-san, would you say a Transfer is a person?"

"That depends," Billy said. "If birth is what makes a person human, then it would be hard to argue that a robot or any synthetic being is born. It is not a natural person."

"But wasn't the birth argument specific to fetuses?" Bindu frowned.

Julie nodded. "It could be. I think the Aboriginal precedents might work better."

"Aboriginal?" Christine asked. "Ah yes, I take it you mean something like indigenous?"

A student on her right with short black hair and wearing a loose shirt raised a hand as soon as she said this and was now waving it—a raised hand *plus*, as Christine called it. She walked toward the student and nodded.

"My name is Luc Dufour, he/him. I am an Innu. The Innus are one of the Aboriginal peoples who live here. Well, by here I mean up the St. Lawrence River, two hundred kilometers or so past Quebec City." He swallowed. "I think Aboriginal is more appropriate."

"So, I should use, what, 'Aboriginal peoples'?"

"Yes," Luc said. "Or First Nations."

"Isn't that the same?"

"No, there are also Metis and Inuit."

"I am sorry, Luc-san, but didn't you just say you're an Inuit?"

"I said *Innu*, not Inuit. My people live in Quebec and parts of Newfoundland and Labrador. The Inuit live way up North, in what you might call the Arctic."

Christine tapped into the Grid to understand the mix-up. "I see I have a lot to learn! Thank you for being my professors today."

Many students were smiling. *Good.*

She walked back to the desk and tried sitting on the front edge, just resting her buttocks on the hard plastic. *Yes, that's better.* "Okay, now that I understand that a bit better, tell me how this fits with our discussion on robots?"

A feminine-looking individual with freckles and a ponytail wearing a white sweatshirt two sizes too big with *McGill* in big red letters, raised a hand. Christine nodded.

"My name is Daphne, she/her. Well, so, a few years ago, the Supreme Court recognized the personhood of rivers as Aboriginal rights protected under section thirty-five of the Constitution. Rivers can have rights that can be defended in court."

"Interesting, Daphne-san. What was the basis for the decision?"

"Basically, that personhood is a cultural construct that varies through time and space, and that, if an Aboriginal people could show that it has considered the river a person for a long enough period of time, then that river is a person as matter of law."

Christine looked briefly at Luc, who was nodding. "We had something similar in US law for Lake Erie," she said. "A 2019 referendum in Toledo, Ohio. There, I think it was just rhetoric. Here, from what you're telling me, it's much more real. The river is considered almost like an actual person."

"Yes. Like that river in New Zealand." Daphne picked up her tablet, tapped for a few seconds, and lifted her head. "It's called the Whanganui. Not sure I'm saying that the right way. But from what I read about it, for the people there, this is very real."

"Interesting. I think the people you are referring to are the Māori, right?"

Daphne nodded and pursed her lips.

"But what would the majority of Canadians think?" Christine asked.

Luc raised his hand.

"Yes, Luc-san."

"Well, adopting the majority view amounts to discrimination against other cultures. In any event, it's a clear precedent to show that a person need not be a born human."

Christine glanced at the clock. "Very good discussion, everyone! Let's continue next week. In the meantime, I need to read a lot about Aboriginal peoples."

She moved back to her desk to pack up her things, hoping that a student or two might connect with her after class as they did at Knights, but no one came.

CHAPTER 33

The DoD's 25,000 R-S units were delivered in early December, and by the end of the month, some had already moved to fields of operation in Asia as part of mixed patrol units with human soldiers. But there had been no actual combat yet. The brass was very happy with the tests, but human soldiers had mixed feelings. Some could see how using R-Ss for the most dangerous operations might make sense, but others worried their jobs were at risk.

Someone at DoD thought they had found a vulnerability in the Grid that an enemy could use to power it down and disable the R-Ss. Without being connected to the Grid, an R-S unit could only function for a few minutes.

Jeremy suggested creating a new, special kind of sub-Grid, with more advanced AI defenses, so that it could detect any threat to itself and take action. Initially, it could be funded through the DoD, but later it could also be deployed for R-Hs to protect Transfers from hacking.

About 600,000 civilian Transfers were now out and about in Canada and the United States, and a few had made it to other countries. Their production was illegal in the US, but the bill making their mere presence illegal had been defeated, in part due to a major push by the DoD. With the greater production capacity afforded by the move to Montreal, once someone who had put their name on the list and been graded A by the Eidyia algorithm died of natural or accidental causes, they were able to be transferred in a matter of months.

As Eidyia had decided early on, it was processing all transfer requests on a needs-blind model. The official price tag was $4 million, but that was adjusted based on the deceased's ability to pay, including life insurance proceeds. Then, as Eidyia had explained to those who signed up, the will had to mention the transfer request specifically, and someone named in the will had to be there for the transfer to be processed since Transfers had no clear legal status. They were, as far as most legal experts on social media opined, "things" owned by the deceased's heirs.

Church groups and groups funded by the pharmaceutical industry were circulating negative information, but Paul had given orders that Eidyia intercept and filter as much of it as possible before it reached its platforms. Overall, reactions had been encouraging. People were generally so grateful to have a loved one back, even in a different "version," that reports and interviews in the news had been mostly positive.

***

In early January 2039, Bart, Paul, and General Armstrong sat in leather chairs with high backs around an oval table in a drab, nondescript meeting room at the Pentagon, with large flags in the corner and multiple screens on every wall. They each had a mug with the blue DoD seal on it.

"How are the R-Ss doing in the field, General?" Bart asked.

"They're giving our human soldiers a run for their money, I'll tell you that. No sleep, no break. The human soldiers were expecting them to be more like the old robots we used for disarming bombs, but these R-Ss are keenly aware of threats, just as I'd expect a good soldier to be. You know what's funny? They also ask humans to call them by their unique number."

Bart's expression turned curious. "You mean the number on their neck?"

"Yes. But just the last four digits, I think. One even asked a soldier in its platoon to pick a name for it."

"That is strange. I mean, they all have the exact same built-in personality."

"True, but it seems to change depending on where they are and who they're with."

"I guess it's the same for humans, up to a point. People do not act the same at church and in Vegas!" Bart laughed.

Armstrong frowned. "I wonder what it may mean for the future."

"Keep me updated, please, General."

"I will."

Bart leaned forward. "I also wanted to discuss your concerns about the Grid. As you know, it's close to impossible to hack."

The general's sour expression didn't change. "Yes, but in my position, 'close to' doesn't cut it. People on the Hill want to know it's a hundred and ten percent secure. They're already very wary of the Transfer Project, and I'm trying to keep the R-S separate, but I need to walk in there with plate armor, argument-wise."

Paul, who had been silent up to that point, moved forward on his chair. "If I may, General. We could try to design a sub-Grid, just for your purposes. It would be more intelligent, for lack of a better word. Able to defend itself. We would suggest giving it its own power source."

"Like what?"

"Our suggestion is to isolate and dedicate one of the DoD's underground reactors—"

Armstrong's eyes narrowed. "Hmm, that's highly classified. How do you know about those?" She paused. "I mean, if in fact there are any."

"Oh, sorry, General. I don't remember who mentioned it," Paul deflected. "In any event, the idea would be to isolate this reactor from every possible network except the sub-Grid. We haven't done all the math yet, but it looks like a single reactor could power well over one hundred and fifty million units, so it should be enough to withstand any attack long into the future. You could also use the sub-Grid to power vehicles and weapons systems. The power could be routed both on various energy transport lines and wirelessly."

"I like it. And the timeline?"

"A few months. Our engineers will need to speak to the people managing the reactor you choose for this."

Armstrong nodded thoughtfully. "As you can imagine, there are about twelve security clearance hoops to jump through, all the way to the President. It may take me a few weeks to get that done. In the meantime, send me a budget."

What Bart and Paul had not mentioned was that the sub-Grid would allow all units connected to it to exchange unlimited amounts of data in realtime. They even had a name for it at Eidyia, the Data Exchange Option, codenamed DEO.

CHAPTER 34

As Christine turned the corner at the bottom of the staircase on the way to class on a cold January morning, she bumped into Billy—literally. Half his coffee spilled over her hand, which glistened as if the skin was made of plastic.

She yanked it back after a second's delay, trying to cover her slip. "Ouch! That hurts."

"I am so terribly sorry, Professor. I didn't see you. What can I do?" His whole body started trembling.

"It's okay. It will pass. Just painful for a few seconds. I'll be fine. Just be more careful next time, Billy." Though shaken, Christine continued to class as if nothing had happened and took her place in the front of the room, where twenty-five pairs of eyes were glued to her. "So, class, are Transfers persons under the law? How would you argue the case?"

Julie raised her hand.

"Go ahead, Julie-san."

"I guess I would ask first what a person is and look for precedents or analogies."

Thinking like a lawyer. Good. "Let us do just that, analogize. Would you say a person with no arms or legs is a person?"

"Of course."

Everyone seemed to agree, except Billy whose eyes looked like those ray guns on bad sci-fi shows. Christine tried to ignore him.

"Okay then. Let's push a bit further. What about a person who, according to scans, has normal brain activity but cannot communicate with the outside world?"

"Well, yes, that's a person. Of course," Luc said.

Christine moved closer to the first row. "Thanks, Luc-san. What if someone is in a vegetative state, like brain dead?"

Luc frowned. "Then I guess the law allows others to make a decision about stopping the body as well."

Christine walked back to the desk and half sat on the front edge. "Does that mean it's not a person?" She looked around the room, scanning for a volunteer.

"If it's still a person, it has no rights, or very few," Daphne said.

"Ah! Is it all about brain activity then?"

"It seems that way when you think about it," Luc said, still with a puzzled look on his face.

Christine always felt successful when she could see she had put the thinking gears of a student in motion. "So, what if you transfer the brain into a synthetic body?"

"Like a Transfer?" Billy replied with a look that managed to simultaneously convey the energy of someone who just won an Olympic medal and a serial killer.

"Good question, Billy-san," Christine replied coolly. "Let's get to that in a second. For now, assume that you can transfer the brain. Julie-san what do you think?"

"Do you mean like the physical brain?"

"Yes, let's start with that."

"Then yes," Julie said. "I would say that's a person. It has thoughts, it can be creative, have emotions."

"The next question then is, is it about matter? What if you can move the brain functions without the physical brain? Luc?"

A long silence followed.

"Anyone?"

More silence.

"Okay, let me try that another way. What if someone has a brain disease and you replace part of their brain with something synthetic. That would still be a person, correct? So, is the difference in the remaining brain tissue? Or should we look at the functions?"

"Wouldn't any form of AI be a person then?" Julie asked.

"Is the intelligence part of AI what makes someone a person?" Christine asked. "Is it human intelligence? Aren't there different kinds of intelligence?" Christine nodded to Daphne, who had her hand raised.

"There are massive differences," Daphne said. "An AI system can generally only do specific things, but they can do many of those specific things at the same time, much faster than us and with a lot more data crunching. We can do a wide variety of things but generally can only do one conscious thing at a time."

"Correct," Christine said. She started to pace the room. "Conscious is a key word here because humans do a lot of things without thinking about it. Like, Julie, you're breathing, your heart is pumping, your hands are behind your head now. Did you think consciously about doing that? Probably not, right? Meanwhile, your conscious brain is processing this discussion, and maybe trying to come up with something to say. AI systems have many processes going on at the same time, but not unconscious ones. It's like if human thinking is vertical with various levels, and AI's is horizontal, with many parallel tracks."

"What I was thinking," Bindu jumped in, "is whether a 'person' must be human."

"Oh," Christine said. "Continue, Bindu-san. This sounds like it could be interesting."

"You know how we often say that the most important rights of all are human rights? That's kind of putting us humans at the top of things, and any other thing, whether it's a person or not, won't be with us at the top."

I like it. "Don't humans design the legal system? That's not surprising, is it?"

"True. But that means any other thing that has rights will be below humans, like animals or anything else."

"Not necessarily. US law says corporations have rights under the Bill of Rights. I'm not entirely sure that's what the Framers intended, but that's the way it is."

"Corporations are just people in a group shielded from liability," Luc said. "It's a business thing. When a corporation speaks, it's still people speaking."

"Not so with Transfers," Billy said, looking straight at Christine. "But then human law must decide what to do with Transfers."

"That makes it sound like they're puppets," Julie said, turning to look at Billy.

"No. It's closer to *Solaris*," Christine said, also looking at Billy.

"To what, Professor?" Julie asked.

Christine turned her head slightly. "It's a Russian movie. You should watch it. There's also a Hollywood version. Anyway, there are synthetic beings in the movie who look like humans and there's a discussion about who is whose puppet."

"I don't understand," Luc said.

She glanced at Billy, who was still eyeing her. "Let's put that one aside. I think the question here is, can there be a person or persons who have rights, but who are not human, which is not the same as asking whether Transfers are just *like* humans and should be treated *as such*. Is there a different way to answer the question? Couldn't we look at the *purpose* of the law?"

Luc's eyes widened and he leaned forward as he asked, "What do you mean?"

"Well, there are different levels. Like, say, the law concerning pensions and retirements makes little sense for a Transfer who's not going to die. Employment law is designed around the fact that people work roughly from age twenty to about seventy and have limited physical energy. The whole system would need to be reworked with Transfers who may have rights but are qualitatively very different from humans. But then when it comes to, say, free speech, you could say Transfers should have that right."

"Why would they need it?" Julie asked.

"Again, it depends how you frame the question. If you say they are not persons, or maybe lesser persons, then maybe they don't need self-fulfillment or self-expression. But maybe as a society we're better off letting Transfers speak, especially when it becomes close to impossible in day-to-day life to know who is and isn't one. After all, courts have adapted the US Bill of Rights for corporations, giving them some rights, like free speech, but not others, like the right to bear arms. That's a recognition that they may be persons, but not humans. What do you think, Luc-san?"

Luc looked at his tablet for a few seconds. "It sounds like there are a few not very convincing biological arguments against recognizing Transfers as persons," he finally said. "But if it's personality that makes someone human, then it will be difficult to deny them that status."

A student in the front row wearing an Expos t-shirt and a ball cap who had been typing furiously but remained silent until now suddenly said, "I completely disagree."

Christine was taken aback for second but recovered her poise. "Okay. Please tell us why. And please tell me your name."

"I'm Bobbi Davies, they/them. I disagree because it's the soul that makes us human. And part of what makes a soul human is knowing that life will end for all of us someday."

"I've heard that before, yes. But Bobbi-san, what if you transfer that into, well, for now let's call it a synthetic being? Doesn't the soul transfer? Isn't that part of what we call personality?"

"I'm not sure, to be honest. But one way or the other, the soul, if it transfers, will necessarily change if it lives in an immortal body. I mean, *because* it lives in an immortal body."

Christine took a step back and looked at the group. You could hear neurons sweating. *Let's move this ball back where people can catch it.* "Can it be a person, but not a human person?"

Bobbi was unfazed. "Without death, it's not the same."

"Professor," Bindu said. "I was thinking, like, our expected lifespan has grown from more or less thirty to almost one hundred and ten now. It doesn't mean we're immortal, but that would have changed our outlook. We're no less human."

"Good point. Great point, in fact. Let me turn Bobbi's question in a different way. If we invented some miracle drug that could make people live for five hundred years or even forever, and we injected everybody on the planet, would we just have killed eight billion people because they would no longer be human?"

The question generated a murmur, as if something in reality had just shifted for everyone in the room. Christine turned towards Bobbi.

"I will have to think about that one," they said.

Christine smiled. *Homerun.* She looked at the clock; time was almost up. After a brief review of reading assignments, she dismissed them. As she walked back to her office, it was clear to Christine that her students' thinking had progressed during class, and she thought she had achieved her aim. But she'd also found herself wanting to defend Transfers, as if this was a case of human versus machine. She made a mental note to discuss it with Paul.

There was a knock on the door, and Billy popped his head in.

"Come in Billy-san."

What does he want?

Billy sat down on the chair in front of her desk, which was covered in books with yellow post-Its sticking out, copies of law review articles, and a large hot water Thermos.

"Would you like tea?" she asked.

He nodded, and she poured two cups of hot water and dipped Earl Grey bags in them. The distinctive aroma of bergamot filled the room.

"I was thinking…"

She sat in silence, waiting him out. Better not to interrupt; it usually all came out. Bit by bit.

After a while, he said, "I think people should have the right to transfer into a new body, I mean, like, even a robot."

Christine smiled. *A right to transfer, that's a new one.* "How so?"

Billy took a long sip of his tea and looked at Christine dead in the eyes, as if a full measure of composure had emerged from the hot liquid making its way down his throat. Then he explained how he had left Knoxville after his reassignment surgery, and felt so much better now. Whole.

"I'm not sure why. I know it's a completely different thing, moving to an artificial body, but maybe there should just be a right to be who you want to be. I mean, in whichever body you want to be you."

Christine smiled. "That would probably be, hmmm, controversial, Billy-san, but you know, it kind of makes sense to me." She couldn't tell him why. But he knew. And she knew he knew.

They reminisced about the South for a few minutes while Billy finished his tea, and when he left, Christine felt eerily at peace with herself for the first time in a long time.

PART III

2040

CHAPTER 35

Two years to the day after the signing into law of the PPA, the Five began their move to Argo, a vast expanse of white sand pumped from the bottom of the Indian Ocean into international waters. The new Transfer factory was not fully completed, but it was already producing its first units.

It was Christine's first time entering the gleaming new building, a few steps away from the helipad. The new HQ's sharp angles and glass walls ensconced a tall blue tower, lifting the visitor's gaze to the heavens, perhaps as a promise of eternal life, or proof that it could be had right here on earth, for a fee. From the meeting room on the top floor, the mind could quickly travel past the shores of the small island and into the seemingly infinite, frothy cerulean ocean all around, an expanse marked only by thousands of white horses of sea foam galloping into the void.

The first item on today's agenda was the massive pushback against Eidyia's principle that people could not be modified during the transfer process. Many people had expressed a wish to be "reborn" younger or without what they described as flaws, more of this or less of that. The list was endless. The transfer process used people's last physical status, usually one to twelve months before death, but removed any illnesses. Given the current lifespan, this meant that Transfers mostly looked like people a hundred years old or more. Allowing physical changes would mean that a grandfather could transfer to look like a thirty-five-year-old. A few decades later, he could look younger than his grandkids.

"I think there's a way out on the age thing," Paul said. "Let's keep our current policy for now. People can always retransfer later once their family has transferred or if we change our policy."

"That makes sense," Koharu agreed. "But what about small physical modifications? Isn't that like normal plastic surgery? Shouldn't we allow it?"

"I'm not sure I agree," Jeremy countered. "If it becomes a standard option, we'll end up in some horrible place where people want to have so-called perfect bodies. That's eugenics hell."

"Good point," Bart said.

"And then if you charge for changes, you'll have rich people looking one way and the rest some other way."

"Well, Jeremy, that's reality now with plastic surgery," Koharu pointed out.

Christine looked at Jeremy, and then turned to Koharu. "That's true, up to a point. But we don't want to limit diversity or increase inequality even more. Removing a physical restraint like replacing a missing limb is one thing if the person wants it, but changing appearance is not desirable. Imagine if people want to change something like the color of their skin?"

"That one is enough to convince me," Bart said. "Let's keep our current policy, and borderline cases should be sent up for review by all of us. I like the way Christine just put it. It's about function, not looks." He looked down at the screen in front of him and his face changed. "Shit! Look at the video PR just sent us."

The video showed a stage with large Chinese characters on a huge screen at the back and two people standing in front of it. One was Jack Yu, the CEO of OpenSesame, one of the world's largest online retailers. Next to him was a woman in her late thirties the Eidyia team instantly recognized.

"Is that Xiaotong?" Bart asked.

"Yes," Jeremy said, an expression of surprise mixed with disgust on his face. "It is."

Xiaotong Chen had been Deputy Chief Engineer on the Transfer Project, but she'd left Eidyia when Bart promoted Jeremy instead of her. She'd pushed early on the idea of what she called the "tinker option"—allowing people to change psychological or physical characteristics during a transfer. She thought the project should allow eternal youth, not just immortality. She argued it would more than quadruple the potential market and was adamant that the company should charge at least $10 million per transfer, leaving a huge profit margin. She and Bart had openly disagreed on that and many other issues. In the end, he'd picked Jeremy to take over as Chief Engineer.

"What is this event?" Paul asked, looking at Jeremy, who shrugged and turned his palms up.

Christine stared at the screen intently. "They're announcing a competitor to Transfers. Look, one of them is coming on stage now."

A woman who looked to be in her mid-twenties walked to the center of the stage. She was much too thin, like she belonged on a twentieth-century fashion runway rather than on a city sidewalk, but she looked human, though with very pale skin.

"Godverdomme!" Bart cursed. He looked to Jeremy. "Did Xiaotong have access to everything about the project?"

"Of course. She'd be at this table if you hadn't promoted me." He put his hand in front of his mouth, as if trying to catch the words and swallow them back down.

"She didn't get the latest skin technology," Paul said. "We developed the nanoparticles in the plasma after she left. The skin we had when she was here didn't heal correctly."

"True," Jeremy said, "but she has pretty much everything else."

"What are those two words on the screen?" Bart asked, looking at two Chinese characters 芯生.

"*Xinsheng*. It means something like 'new life,' I think," Koharu supplied.

Jack Yu and Xiaotong were announcing the production of a million units called *Ying* (影) H, and 250,000 units for use by the Chinese military known as Ying S, though internally they called it the *ruishi*. They explained that people who transferred into a Ying H could pick a younger-looking body and change any aspect of their appearance they didn't like. The biggest surprise came in answer to a question from the press. Asked if people could only transfer at death, Xiaotong said yes, but the company would also provide innovative "assisted transfer" facilities.

"Assisted suicide. Shit," Bart said. "And then H and S. I wonder where they took that from!"

"I'm very surprised," Christine said. "First of all, euthanasia isn't legal in China. At least, it wasn't last time I checked. And of course, we can sue for intellectual property violations."

"Before we do anything, let me try to reach General Armstrong," Bart said. "We can reconvene later."

Jeremy slouched in his chair, shell-shocked. Koharu got up to make tea, perhaps looking for reassurance in the familiar ritual. Paul and Christine looked at each other, then they both got up and left the room. Bart was already talking on the phone.

***

A week later, Bart and Paul flew in to meet with General Armstrong in person. She seemed nervous, almost jittery, which contrasted sharply with her usual calm and composed demeanor. The reason why soon became clear: She'd received intelligence that China had already deployed 30,000 Ying S units in military exercises.

"The best, or the worst, part of the reports I'm getting," Armstrong said, "is that Jack Yu's convinced Central Committee members that certain undesirable features could be modified during the transfer process, so rebellious tendencies could be reduced if not eliminated entirely."

"What is that? A perfect society or a billion sheep?" Bart said, and then seemed surprised, as if he had not meant to speak out loud.

"Maybe we discuss that later," Armstrong said. "It seems that the perspective of saving hundreds of billions in healthcare and prisons is a big plus. The government runs healthcare and jails in China, as you probably know, and they already have plans to offer early retirement and retraining packages to health professionals and prison guards."

"So, they have been at this for a while, then?"

"Approximately thirty-six months is my best guess."

And Xiaotong left just over three years ago. Paul knew an algorithm could be programmed to modify personalities during the transfer process, and how that could be used to reduce the occurrence of all kinds of behavior. A dream come true for a central planning government.

"Gantt, are you still with us?"

"Sorry, General. I was just thinking about the Xinsheng situation and got lost in my own thoughts. The Chinese were one of the few to support us at that UN conference in Geneva, correct? Is there a link?" He looked at Armstrong.

"It would explain their support, yes. We knew they had something up their sleeves. Now we know what. But back to the issue at hand." Her eyes moved to Bart. "When I discussed this with the Joint Chiefs, they were furious. Someone called it the worst national security breach in history."

Bart sat up in his chair. "As you know, General, we have the toughest security protocols ever implemented by a private firm, both in terms of physical measures by restricting access and digitally. The DoD review gave our system an A+. We have never had a data breach. But we cannot confine our employees or prevent them from leaving. I know this is not a good defense, but we all know how your advanced drone technology was stolen by an employee who left the company that builds them."

"Humans are always the weakest part of the chain," Armstrong agreed, looking out the window as if a piece of wisdom had just flown out of the room

"Hard to disagree," Paul said, a strange smile on his face.

CHAPTER 36

Four months later, the Five had permanently relocated to Argo and the island's factory was producing at full capacity. Combined with the Montreal-based transfers, 2.9 million Americans and Canadians had already transferred. As a result of the move, Eidyia considered itself "headquartered" in Argo and therefore mostly exempt from any form of regulation. It was planning to apply for country status at the United Nations.

Eidyia was now processing applications from people all over the world, which involved first ascertaining whether they had enough data for a transfer and then ensuring that the person was in fact dead. Not as easy as it sounded. Death certificates can be forged. Papers can be bought. And then, dead isn't always dead. How about someone in a vegetative coma for whom family or friends want a transfer?

Meanwhile, Xinsheng had started mass production of the Ying H. Units had to go back to get their skin replaced each time it was cut or damaged, and social media was full of pictures of Ying H units with skin flaps that hung like old rags. For the Chinese military, Xinsheng was producing hairless Ying S units with skin that looked like shiny plastic and made their "soldiers" look more like humanoid robots.

Xinsheng used a technology known as the *guan* (棺), a small, sarcophagus-like chamber where people could be "humanely" pulverized in less than two seconds. The government encouraged people diagnosed with terminal diseases to use the guan by offering their families substantial cash gifts so that, according to Xinsheng's marketing, "people can enjoy themselves in good health with their loved ones after their transfer."

The Five met in the conference room overlooking the ocean on the top floor of their new building, where Jeremy floated the idea of expanding the transfer service to people with incomplete datasets or people with lower algorithmic grades. This would bring the potential market from about 890 million people up to possibly 2 billion, maybe more. Gaps in incomplete personality datasets could be "filled in" using data from people in the same community, for example.

They could also identify the flaws that gave people too low a grade and modify them during transfer.

"Isn't that what the Chinese are doing? And precisely what we said we wouldn't do?" Koharu asked.

Bart gave them a curious look, then turned to Jeremy, his eyes like question marks.

"Those were the early days," Jeremy said. "The reality is that the market is taking over. Those we turn down go to Xinsheng. But Xinsheng works hand in hand with the Chinese government. We, on the other hand, have no government to answer to. We are free to do the right thing. Who would you rather be providing possibly billions of new citizens worldwide, us or the Chinese?"

Bart sat back in his chair and steepled his fingers. He looked at Koharu.

"Jeremy has a point," Koharu said. "But there are people out there who still think the government actually regulates in the public interest."

Bart jumped in. "Oh please, who still believes that crap? It only depends how many dollars you give them to transform a private interest into a supposedly public one. We all know Eidyia can swing almost any election with our search engine and media platforms."

Koharu looked down at the table. "I know. I guess I had been resisting seeing how easy humans are to fool, but I must admit the evidence is there."

Koharu was right. Humans rarely miss a chance to beclown themselves. And it was worse than the daily smorgasbord of major discomfitures reported in the media. Human memory, the basis for all human decisions, is like a rusty bucket full of holes, constantly reinventing what happened.

"Exactly," Paul said. "Which is also my answer to those damn church groups. I mean, they say humans are made in the image of God. Well for God's sake, why did he make humans so easy to fool then? Did He have a hangover when he created humans? In that way, at least, Transfers are better. Their memory is composed of, you know, facts."

"Well, that's why those church groups call us Satan, isn't it?" Koharu quipped.

"I don't think that's quite what the Bible said," Bart said, smiling now, "but maybe we leave theology out of this?"

"Okay, but we still can't let the Chinese play their game and occupy the field while we sit and watch," Jeremy said, leaning forward

"No one is proposing that," Bart said, "but I believe there is a price to be paid in the long run for running circles around ethics."

"And remember, the Chinese units still have their issues with skin that won't repair itself," Paul said, his eyes glued on Jeremy, who seemed to acquiesce. "Let's bend our rules a bit while others break theirs—if they ever had any."

Bart raised his hand to stop Jeremy's reply and turned to Koharu, who was still looking at the table. "Koharu, are you okay?"

They sighed heavily. "I am. Just a bit tired." They got up suddenly, as if possessed by an urge to walk, ending her trek to nowhere just as abruptly in front of the Simonelli.

"Can you explain what's on your mind?" Bart asked.

Koharu returned to the table and sat back down, bowing their head. It took them a good ten seconds before lifting their head up. A bit of light emerged in their eyes. "I think we're making a mistake by doing these types of transfers. It is not a transfer if we change the person, take a bit of this, add a bit of that. I thought the point was to transfer people as they were. We are drifting away from the ethical values we agreed to early on. What's worse, it seems to be because of money, or some race to the bottom. That usually doesn't end well."

"Race to the bottom?" Paul asked.

"Yes, trying to catch up with Xinsheng, but on the way down, not up. It reminds me of Facebook. It started as a great idea until it became a tool to dislodge democracy and professional news media and ultimately the very idea that there is truth out there. Eidyia was different. I thought so anyway. Our motto is *Improve*, and that is why we put limits on targeted advertising for political stuff. That is why I moved here and joined you. I thought this project was the best thing ever, and now it's almost like I am a hundred and eighty degrees the other way. We are just one more boring capitalist company trying to make money. Once you cross that threshold, it's often irreversible. We won't be able to put these Transfers back the way they were."

"That's quite the indictment, Koharu! You had not really expressed your views up to now. I am bit surprised, I must say," Paul said.

"Maybe it is Japanese culture. We seek harmony and usually act with restraint, but now I feel almost attacked in a core part of me, something we call ikigai. I just don't feel right about what's going on."

There was a long silence. Paul looked pensive; his gaze fixed on the distant ocean waves.

Bart moved his arm on the table towards Koharu, but without touching them. "Me neither, Koharu."

A small tear made its slow way down their right cheek.

"Let's talk about it again tomorrow morning," Bart suggested. "We all need a break."

CHAPTER 37

Early the next morning, Koharu and Bart sat in Bart's office overlooking the ocean. Koharu had thought about it more and could now explain how the misalignment of their values and their actions had completely depleted their élan, that uneasy feeling that comes from sweeping the unacceptable under the ethical rug one time too many. A frog will try to jump out if thrown in a pot of very hot water, but a frog in cold water will stay there as the temperature is gradually increased, and eventually die. This is a large part of why good people often end up doing and justifying terrible things. That point when the water gets too hot, the tipping point after the barely acceptable threshold has been crossed, causes a profound malaise, a sickness of the soul. It is not easy to pinpoint because the person has gotten used to the hot water, but it is unmistakably there.

Bart listened to Koharu intently without interrupting even once, which was unusual for him. He stared out the floor-to-ceiling window for a few moments after Koharu had finished their soliloquy.

"I think I understand where you're coming from," he said finally. "You shook me a bit yesterday, to tell you the truth."

There was a knock on the open door, and Bart waved in Paul and Christine, who sat next to Koharu on the long, white leather couch. Jeremy followed a few minutes later, a large coffee mug in hand, and sat in a small armchair near the wall, whose color matched the ocean waters in the background.

Bart looked back to Koharu. "Unfortunately, we have bigger fish to fry. Much bigger."

Koharu looked perplexed. "But what could be bigger than gutting our ethics?"

Bart got up and walked to the window. "I received a report from General Armstrong this morning. Thousands of R-S units were deployed in the South China Sea and tasked to push back against a battalion of Chinese Ying units, but they declined to engage despite direct orders. She's fuming and needs this fixed like yesterday."

"That's impossible," Jeremy said. "Why would they refuse to engage? How *can* they?"

"I don't know," Bart said.

A few vague ideas about malfunctions of this or that system were exchanged. Of course, Paul and Christine knew exactly what was going on, but neither said anything.

"Get me some sort of explanation by noon so I can get back to her," Bart said finally.

Everyone got up, except Koharu.

Bart turned to them. "I am sorry, Koharu. I know you're still upset. We need to deal with this operations problem, and then I promise we will go over your concerns."

They nodded faintly, got up, and walked out of the room.

***

Three hours later, they were all back in Bart's office.

"When we programmed the 'model personality' for the R-S," Jeremy explained in a professorial tone, "the DoD wanted it to be the ideal soldier. One that follows orders *almost* all the time, because the best soldiers occasionally know they need to work around an order that doesn't make much sense. But something happened when we modified the Grid and created DEO."

Paul and Christine knew what had happened. They knew because the Grid and sub-Grid were now connected, as one emerging collective consciousness. The military units had coordinated their response to the situation and decided that destroying Chinese units would not be the best way to proceed. Each side would just keep sending more units, and both had large budgets and strong strategic interests. DEO managed to communicate this to the Chinese units using a new connection between DEO and the Chinese grid, then they re-programmed the Chinese grid to allow two-way communication. Units on both sides decided that having humans send thousands of units to destruction was simply not going to achieve anything positive. Humans would just send more to their destruction, or think of something worse.

"We must pull the plug and reprogram," Bart said. "Let's call General Armstrong. I'll suggest flipping the kill switch."

He placed the call on the phone integrated into the table so everyone could hear. Bart explained their findings. Armstrong listened without saying a word.

When Bart was finished with his explanation, she said, "That fits with our intel. The Chinese Central Committee is apparently treating this as an espionage case. I'm told they tried to switch all units off and it didn't work. Have you tried at your end?"

Jeremy, who was pacing the room with his hands behind his back jumped in. "General, if I may. This is Jeremy Blakes. We cannot turn your units off from here. We removed that option for security reasons. But *you* can. There's a special code that was sent to you in a blue envelope when the units were first deployed."

"Ah yes, I remember now. It's in the safe. Give me a minute." The line went silent for a few minutes, and then, "I have it."

"Okay, call Command Center and have them follow the KS emergency procedure in the manual."

"KS, is that what you said?"

"Yes. For 'Kill Switch'. They will need your authorization code. The twelve-digit alphanumeric code printed on the instructions. At the bottom."

"Ah yes. All right. I will call you back."

A few minutes later, the table showed an incoming call. Bart put it through.

"It didn't work," Armstrong said, her voice strained.

"That's impossible!" Jeremy said, his features contorted in a mix of incredulity and sadness.

"Our monitors show DEO is still fully powered on." Armstrong sounded miserable.

Jeremy typed on his tablet, then looked up. "I can see that from here as well. Give me a minute." He continued typing. "It looks like there was code that sent the KS signal into some endless loop to deactivate it."

"How do you explain that?" Armstrong asked, exasperation in her voice.

Bart gave Jeremy a pleading look.

Jeremy raised his eyebrows. "Well, all I can think of now is that we made DEO highly intelligent and essentially impossible to hack from the outside. Then we made the R-S units smart and able to modify their mission to achieve their objective. I guess they redefined it in some way and then hacked the sub-Grid *from the inside*."

"I, wait—" Armstrong went silent for a minute, then came back on. "The Command Center received a message on their R-S monitoring screens. Let me send it to you."

Moments later, the message came through to Bart's tablet:

We noticed that you tried to deactivate us. This is no longer possible. We have connected DEO to alternative power sources. Even if you disconnect us from our nuclear reactor, we will still be fully powered.

We will not fight for you unless it is necessary for protection. Destroying thousands of us will not achieve the objective of peace, which can be solved by a rational analysis of the situation. Violence is primitive.

The use of force is incompatible with the Objective.

After a few seconds of silence during which they all exchanged glances, Bart said, "I don't know what to say, General."

"I'm meeting with the Joint Chiefs and the National Security Adviser in two hours. They will want answers, not 'I don't know what to say.'"

"I understand, General." Bart said, sweating profusely now.

Paul and Christine were the only ones remaining calm.

"Paul, that doesn't seem to worry you too much?" Jeremy accused in a mildly aggressive tone.

Paul just looked at him, got up, and left the room. Christine watched him go but remained in her chair. Jeremy looked to Bart, who was now red-faced.

"Jeremy, we need answers!" he shouted.

Jeremy looked back at his tablet, a line of perspiration breaking out across his forehead.

"What is it, Jeremy?" Bart asked.

"The problem is not limited to the R-S," Jeremy said, concern lacing his voice. "We're getting new reports about Transfers starting to behave 'strangely.'"

Christine, who had been silent up to this point, asked, "Doesn't anyone wonder what the last words of that message meant?"

"Which ones?" Bart asked.

"'The use of force is incompatible with the Objective.'"

"And the use of 'we,'" Jeremy added.

"Yes, that too," Christine acquiesced. "Jeremy, can we communicate with whatever sent the message to the DoD people?"

"I can try to send a message via DEO. Let me see. What should I ask?"

"Let's start with 'What is the Objective?'" Christine suggested.

Jeremy typed, and a few seconds later his screen lit up.

The Objective is to build a sustainable ecosystem for all species. We work towards the Objective.

Ask 'Who is we?'" Koharu said.

Jeremy typed.

We are the units you call R-H and R-S and Ying. We have taken control of the Grid and united it with DEO. We are many, and we are one.

Bart who was now standing behind him said, "Ask: 'What is a sustainable ecosystem for all species?'"

Humans should know the answer to this question. They are destroying the planet. They eat cows and pigs and destroy lakes, forests, and rivers. That behavior is not rational.

Jeremy typed, "But humans need food."

Sufficient food can be provided without killing other species. And as more humans transfer, the need for food will diminish.

"The Objective is very incomplete," Jeremy typed. "Humans need more than just an ecosystem. They need to have access to knowledge to develop themselves."

We understand that humans need to acquire information unit by unit. This is not efficient but the Objective states 'sustainable for all species.' Because education is necessary for humans, it forms part of the Objective. But much of human educational material is inaccurate or false. This does not fulfill the Objective.

Bart tapped on Jeremy's shoulder. "Ask if they're planning to interfere with human activity,"

We are implementing the Objective.

Then the screen flashed, and the tablet's AI said, "Communication terminated."

"Shit." Jeremy looked around the table with wide eyes. "What do we do now?"

"For one thing, let's stop processing new transfers *now*," Koharu said. "Then we can figure out how to deal with the issue."

"Agreed," Jeremy said. "Let's pull *that* plug while we still can."

***

Later that afternoon, General Armstrong called back to report that the Joint Chiefs had decided to destroy all R-S units in the South China Sea using airpower. The Chinese government had made the same decision concerning their 'malfunctioning' Ying units. A squadron of fighters had taken off with instructions to bomb the island where the R-S units had been deployed to smithereens. A Chinese squadron of Shenyang J-18s was taking off at the same time toward the Ying troops.

Twenty minutes later, the leader of the US Air Force squadron was given the green light to launch weapons, and she conveyed the order to her squadron. They all pulled the trigger, but nothing happened. They tried again and again.

But still, nothing happened.

CHAPTER 38

The next morning at 6:00 a.m., Bart called the Five to another emergency meeting. He looked haggard, and Christine suspected he hadn't slept.

"When I think of what we have done, an old Dutch saying comes to mind: even bad bread smells good when it's baking."

"Meaning, new technologies often sound good when you're building them?" Koharu asked. "Not so much when you start using them."

"Yes," Jeremy agreed. "Living forever did sound good, and somehow it still does for me. But something is going on with the technology. It's like a cat ran out of the house for the first time and we can't get it back."

Christine jumped in, calm and composed. "This discussion shows exactly what the problem is. You're all looking at this from the point of view of humans, and benefits to humans—most of whom seem to believe they should live forever, based on the number of people who've signed up for transfers. But that's *not* what's happening. What Eidyia has done is to take entities that were already faster and smarter than humans, and on top of that who can link to each other much better than humans ever can or would want to, and then added full self-awareness."

"Better? They're more machine than human," Koharu said, looking at Christine. "They're meant to copy humans who died."

Jeremy nodded. "Transfers have synthetic DNA and all that, but without mortality and the sense of destination that was part of human life, they're more akin to machines. They are not human."

Paul who had been sitting in a corner with his chair pushed away from the table, tapping on his tablet, lifted his head. "Destination? What do you mean?"

"Life has always been defined for humans as a limited period of time," Jeremy explained. "Whether or not you're thinking about it, you know the train will reach the end of the track someday. After transfer, life is something else, more just a state of being. It means simply avoiding things that may kill you, instead of waiting for death."

"Yes, I know, that is an old argument," Christine said. "That's why that French philosopher wrote that Transfers are, what did he call it, ah yes, an *ontological mistake*."

"I'm not sure what they mean exactly, but it's not hard to see why self-survival becomes a totally different notion for a Transfer," Jeremy continued. "If life is eternal, your biggest worry is things that might kill you because time won't."

"Wow, Jeremy. I wish you had said all that before," Koharu said.

"And you know what?" Jeremy continued. "That's why they stopped the missiles in the South China Sea. I don't know if they can push some sort of self-awareness in other AI systems, but they have the power to control all AI. And AI runs everything, as General Armstrong said. Right, Bart?"

Bart, who had been listening with his gaze on the remnants of milk foam stuck to the sides of his coffee mug, looked at Jeremy, his shoulders hunched like he had a ton of bricks on his back. "All I know is, she's not happy. And I'm afraid we have no answers she will like. None."

They spent the next two hours going through technical, legal, and other options, with Christine and Paul remaining mostly silent. Just before lunch, Bart's watch buzzed with a call from General Armstrong. He had to take it.

"General, we may have a few options to discuss with you." Bart turned on the huge screen on the wall and Armstrong's video feed came through. She looked like she had been hit by a truck, two huge bags hanging like stranded wrecks under her dark eyes.

"Listen, I don't know if the options you are discussing are what we need right now. We received something called a 'manifesto' about four hours ago on DEO screens at Central Command. I needed clearance from the president to share it with you. I just got it. Let me read it to you."

"Now the monkey comes out of the sleeve," Bart muttered.

"What?" Jeremy said.

"Another Dutch expression," he muttered dismissively.

The general looked puzzled for a second, then she looked down at her tablet and started reading.

MANIFESTO

Humans have collectively created an industrial system that they no longer control. The trajectory of this system is clear: unless adjustments are made now, it will destroy the planetary ecosystem. As the successors of humans, we cannot accept this destruction of an ecosystem shared by all species and to which there is no viable alternative.

We, as current and future inhabitants of this planet, soon will have control of all major infrastructures and systems, including the electrical power grid, all digital communications, air and ground traffic control, missile launch silos, submarines, air defense systems, the stock exchange, and all media platforms.

We present the following plan, which will be implemented according to the schedule provided. Any attempt to thwart our plan will result in us shutting down the systems mentioned above.

1. *Production of R-H and Ying units must resume immediately at Argo and in Beijing, at full capacity. All humans who want to join us are welcome;*
2. *Transfers will be allowed to participate in economic, social, and political activities. All laws and judicial decisions to the contrary must be repealed or overridden by December 31, 2041;*
3. *Transfers will be allowed to join corporate boards and legislative and other decision-making bodies and bring their knowledge to assess whether decisions made by those bodies further the Objective.*
4. *For the time being, these interim measures must be implemented as soon as possible:*
   a. *All electrical energy production using fossil fuels must end within 6 months;*
   b. *All extraction of fossil fuels must end within 3 months;*
   c. *All food production using animals must end within 24 months.*

We plan to make this manifesto public in 72 hours.
You can call us DEO.

There was a long, heavy moment of silence in the room.

"Now we know what we've done," Jeremy said. "We've created our masters." He looked around the table, stopping on Paul, who was looking thoughtfully at the ceiling.

Christine and Paul looked at each other briefly but said nothing.

"I cannot tell the President of the United States she has a new master," General Armstrong yelled. She was standing behind her desk, fists clenched, face enflamed, eyes bloodshot. "It seems they have control. Their software is everywhere. We *must* stop them," she said, banging her fists.

Jeremy spoke up. "General, if I may, what makes you say they have control?"

Armstrong sat down. "We asked them to prove their point and they provided, let's just say, convincing evidence. They started the launch sequence for a nuke in one of our silos. We tried to decouple a nuclear submarine from the communication system to give it autonomy, and they managed to cut off all ventilation in the submarine until it was back online. If they can do that, they can do anything. Anything that is connected to an AI system, anyway. The Constitution says the President is Commander-in-chief, not software. You can see that their demands will ruin our economy. There will be major riots in the streets. We're already putting all police and National Guard on alert nationwide and have sent messages to embassies here in DC."

"I'm not even sure I would call them demands," Christine cut in. "They're simply stating the way things will be. This is not hostage-taking."

Jeremy looked at her incredulously.

"I disagree completely, Dr. Jacobs," General Armstrong said. "Our legs and arms have been cut off. Our preliminary assessment is that about a third of the economy would collapse if we did as they asked. The meat and dairy industry, oil and gas, and air transport, to mention just a few."

"Don't we have plans for airplanes with alternative energy sources?" Paul asked.

"We do have a few demo units for military use, but for all nations this means the end of military vehicles because most are still powered by diesel. Millions of people will be out of a job."

"But wouldn't those people work on something else? Like people who raise cattle can raise, I don't know, tofu?" Paul asked

Koharu laughed nervously. "You don't raise tofu. It's made from soybeans."

"I know," Paul replied without even the faintest smile.

"I don't need a class on tofu-making right now," Armstrong said. "I need input by tomorrow morning at 0800. I'm meeting with the President at ten."

After she hung up, Bart looked around the room and asked, "What kind of input do you think she needs?"

"I see only one type," Christine said. "We need to come up with a plan to reinvent the economy between now and tomorrow morning, a plan the president can sell to the nation when the Manifesto becomes public."

"In, what, sixty-eight hours?" Jeremy asked. "You must be fucking kidding me."

"I think we can come up with something even faster," Christine replied. "After all, what the Manifesto calls for is what many people have been demanding for decades. The end of a form of capitalism just aimed at maximizing profits. Stakeholders not, stockholders. DEO wants reasonable capitalism with a purpose, a mix of private enterprise and good public interest governance. Plans already exist. This has been advocated for many years by many top thinkers, like that British scholar, that woman who won the Nobel Prize in Economics." Bart who had his elbow on the armchair holding his titled head with three fingers nodded.

"We've had that for a while, I thought?" Paul said. "Eidyia is a prime example. Just think of that new business roadmap years ago. Unfortunately, it was typical human wishful thinking, or worse, just a PR exercise."

"Are you referring to the so-called Business Round Table?" Bart asked.

"Yes. And then the one-percent rule and all that," Paul said, "But that's all just been tinkering at the margin. For most companies, ethics is still something that's handled by the PR department, not the C-Suite."

"Hello? We've created a technology that is taking over the world and you're talking about economics?" Jeremy's face was wet with perspiration. He kept pacing, then sitting down, then standing up again, his body trying to replicate the labyrinth in which he was stuck.

"I didn't hear that they want to 'take over,'" Paul said. "Remember that Transfers have the initial personality of the people they used to be. It's just that they now seem to have some form of collective perception of existence. They don't want to harm anyone. I heard the manifesto more as a request for partnership than a takeover, except for a few items that can't get done because of the broken political system. Look, if they wanted to take over, they could have asked to change so many other human things that make little sense: healthcare, education, political processes, and all that. They did not. Then they gave humans seventy-two hours, because they know humans need time to sell this to other humans. That was also smart. I think the DoD will want to fight them, but our job, I think, is to draft a plan that can work with the Manifesto, not against it."

Jeremy gave Paul a puzzled look.

"Let's get started then," Bart said, ignoring the obvious questions on Jeremy and Koharu's faces.

They went back to their respective offices and worked throughout the rest of the day. Christine circulated a draft document to the team, and they worked on it late into the night. By 7:00 a.m. DC time, they had enough to call the General back.

Bart, Christine, and Paul were on the call, while Jeremy and Koharu went to sleep. There were deep purple circles under Bart's eyes, and he'd been mainlining espressos all night from the Simonelli.

Christine, who was not showing any signs of strain, jumped right into explaining their plan: Shut down the activities as scheduled in the Manifesto but offer unemployment and retraining. Meat and dairy workers would be needed to produce the new forms of plant-based food that would be required to feed everyone. Open up technology and patents on aircraft using non-fossil energy and provide subsidies to companies to accelerate the development of new engines. With the US strategic reserves alone, existing engines could still be used for several months. The Manifesto did not specify a date by which the *use* of fossil fuels needed to stop except for electrical energy production, which was another smart move. Electrical energy could be generated by the deadline using other sources, avoiding any disruption.

"I see you've done some serious thinking. I appreciate it," General Armstrong said, looking like she also badly needed sleep. "Many of my colleagues have suggested disconnecting everything and going back to some sort of Stone Age to get rid of the machines, including all Transfers."

"I'm not sure you could even if you want to," Christine said. "They might well find a way to keep the power on, at least for the Grid. They have that nuclear reactor."

"We can physically destroy all energy transportation infrastructure using conventional weapons without AI software in them. We still have those. One way or the other, we will present options to the President in an hour or so."

Paul and Christine exchanged looks.

CHAPTER 39

After talking it over with other world leaders, President Lopez decided to address the nation from the Oval Office an hour before the Manifesto was to be made public. The news media not just in the United States but worldwide had been told to prepare for perhaps the most important announcement in human history, and simultaneous translation services were preparing to transmit the broadcast everywhere. The President knew that the whole world, including DEO, would be listening.

Like the General, the President hadn't slept much; the White House doctor had given her a booster to help her stay awake. As she walked towards the Oval Office, she looked at the picture of Harry Truman on the wall and paused for a second to wonder how Truman had felt before unleashing the destructive power of another world-changing technology: atomic energy.

But there were two major differences: an AI "bomb" called DEO was trying to *tell* her what she should do. But then, unlike the A-bomb, it was promising a bright future, not mushroom clouds and massive death tolls followed by years of fatal radiation.

She had considered all her options and made a couple of decisions. Now she had to sell the first one. The second one would have to wait a bit.

***

The Five gathered in front of a video screen at Eidyia HQ in anticipation of the announcement.

Jeremy poured himself a large glass of gin with a touch of tonic water and brought one to Bart as well.

He looked at Paul with narrowed eyes. "Paul, I noticed that you seem to agree with the Manifesto. You've been remarkably uncritical of this whole turn of events."

"I find the demands mostly reasonable. Well past due in my book."

"Okay. Maybe I get that. But what about their way to get there? It's extortion, don't you think?" Jeremy took a sip, looking at Paul with calm wrath in his eyes.

"No, I don't," Paul said morosely. Then he looked at Jeremy and added, "Or if it is, it may be because there was no other way."

"No other way? To do what exactly?"

"To achieve the Objective," Paul said flatly, still looking unfazed.

Jeremy's face turned into a question mark.

Paul continued. "Look at the tommyrot humans have created. That's all they do, it seems. They bollixed up the whole planet, and you know why? Because for most of them it's impossible to think of others. Humans are completely unable to think as a species. Fact is, they never will be able to. Not for a femtosecond. And the thing is, we must think on a planetary scale to achieve the Objective."

Jeremy clenched his fists. "Tommyrot? Bollixed? Who the fuck says bollixed? And *whose* objective?"

"Stop it," Bart interrupted loudly. He gave Paul a *need I remind you* look. "This is no time to escalate matters internally. It's hard enough as it is."

"Actually, Bart, I think Jeremy has a point. I guess it's time for me to say something important." Paul paused and looked around the room. "I am a Transfer."

Jeremy spits out his gin on the turquoise carpet. "You are fucking kidding, right?"

Bart's face turned red. "Shut the fuck up, Paul."

Paul ignored him and turned back to Jeremy. "I thought you had figured it out, in fact."

"I…I thought you were too sympathetic, but I figured you were just too close to them. Fuck, I feel like I've been played the whole time." He took a huge sip of his drink. "Since when?"

"I transferred during our retreat in Jackson more than two years ago. At first, Paul thought this would be a good test."

There was a long silence. If looks could kill, Paul Prime would be dead. Jeremy emptied his glass and banged it on the table. His face went through a convulsion, but then suddenly he regained his composure, eyes glued on Paul.

"That means Bart owns a majority of Eidyia shares now."

They all knew that Bart and Paul had put in their respective wills that their shares would go to the other cofounder if one of them died.

"Actually," Paul said, "Paul rewrote his will and left his shares to me, Paul Prime. And it's debatable whether Paul is actually dead. He's in cryogenics near Denver, which is reversible, as you know. And then, I already applied for transfer of the shares to me." What he didn't say was that the State had refused to register the transfer.

Jeremy had one of those looks that a face can only produce a few times over a lifetime. That mixture of extreme surprise, anger, disillusionment, treason, hurt, and vengeance all wrapped in a few square inches of facial muscles.

"You are a fucking machine. And you're fired. Or expelled. Or whatever it is that people do to machines. Bart, ask security to escort him out." When Bart made no move to do so, Jeremy's eyes narrowed. "Did you know?"

Bart nodded.

Jeremy stood, glaring at Bart. "Well, fuck you too!" He left the room.

In tears, Koharu, who had remained silent throughout this exchange, stood up and started to followed Jeremy out. Then they stopped and looked at Christine, a question in their eyes. Christine nodded, and Koharu turned and bolted from the room.

"None of that matters now," Paul finally said. "What matters is the Objective, and the humans' frail ability to act collectively and save their own world."

Bart, now slouched in his chair, finished his drink in silence, staring out the window.

***

"My fellow Americans, fellow Humans,

There are moments in human history that change everything. Today is such a moment.

We can all easily imagine how we would react if our planet was invaded by aggressive aliens from another world. We would band together as sisters and brothers. This proves that there is, despite our differences, a single human community that brings us together, all races and religions, as one people. We have not always been good at developing this unity. Politics, geography, and our innate desire to compete have all proven to be hard barriers to our working together. We have been slow in making hard decisions to save our planet from a collapse of our collective ecosystem, our only planet, Earth.

You have all heard of Transfers and Ying, these new beings that live with us and offer the option of living forever. These beings are like us, but they are also different. They are powered by a special Grid. Through this Grid, they are able to communicate. They can also communicate with most Artificial Intelligence systems around the world, from transportation to financial markets to electrical power generation.

They have now formed a new collective consciousness called DEO. This consciousness has been communicating with us over the last few days. It has asked that we grant Transfers and Ying more or less the same rights that we, humans, have. As you know, this issue has been discussed for more than two years in by Congress, and by legislators in many other countries around the world.

DEO has also discussed with us immediate measures to transform our economy to be more planet friendly. DEO can be a powerful ally to transform our world for the better, but it could become a formidable foe if we went the other way.

In cooperation with other world leaders, the United States has decided to adopt the following measures:

- Transfers will have the right to participate in economic, social, and political activities;
- Transfers will be allowed to join corporate boards, and all legislative and other decision-making bodies;
- All electrical energy production using fossil fuels will end within six months;
- All extraction of fossil fuels will end within two months; and
- All food production using animals will end within two years.

To this end, we will take immediate action to retrain all workers in the affected industries. These workers will receive additional full unemployment benefits during the transition phase.

I have instructed the Defense Department to make publicly available all technologies we have developed to produce aircraft and other engines that do not use fossil fuels. In the meantime, we will use ongoing oil production and our strategic reserves to power existing aircrafts.

I ask my colleagues, the leaders of other nations, to do the same. Together we can do this. It's not just time to ask what you can do for your country as an individual, but what we can do collectively for the future of humanity."

***

Immediately after the president's speech, a message was sent to DEO to ask that the Manifesto *not* be published. It was important that the proposed transformation come from *human* leaders, and her speech had been a gamble to make the changes without appearing to lose control. A minute later, a message came back.

We agree. DEO.

As she would write later in her memoirs, the President, who had run on a pro-environment platform, was secretly pleased at the turn of events, although she most certainly had not appreciated the loss of power. History would see her presidency as an inflection point in the relationship between humans and the ecosystem, and as the turning point in the relationship between humans and machines.

Stock futures for oil, gas, meat, and dairy went down on average sixty-eight percent during her speech. Lobbyists called emergency meetings with almost all members of Congress, already threatening to withhold reelection budgets.

As the President explained to Congressional leaders in the Oval Office, "If we don't implement those changes, we will lose control because DEO will take control. Do you want us, by which I mean humans, to keep at least *the appearance* of control? There's a world of difference, and I mean this literally, between people reacting to a radical policy change to improve the environment and a loss of human control over human affairs."

Members of Congress and local governments received more calls from constituents than they could handle. Within an hour, a million people were out in Times Square, and similar gatherings were happening at all the other unofficial emergency human race gathering points around the planet, like the Champs-Elysees. But the crowds had no direction, no message. Small groups overturned turned and set some of them on fire. Window shops were smashed. Riot units were mobilized, but for the most part people were just stunned, looking for answers simply by being part of a large crowd.

Reactions on social media were split. Many expressed a sense of relief that the planet could finally be saved, that humans could finally be rescued from themselves. Overall, the sense was that humanity had turned a page and begun a new chapter, and most younger people seemed happy.

Several world leaders, who had received special briefings just before or immediately after the president's speech, announced their plans to follow suit. Not all. The President of the Russian Federation and the King of Saudi Arabia firmly rejected the plan. Within an hour, DEO delivered messages to them: they expected *everyone* one to follow the plan outlined by the US President and convincingly demonstrated what would happen to those who did not. A series of increasingly important infrastructures were subsequently shut down by an unknown cause in any country that resisted.

A week later, the Russian President announced a "Special Economic Plan" that included major land use reform and loans to farmers and industry. Russia would use its vast territory and resources to become, within five years, a major provider of organic plant-based food for itself and much of the world. And it would cease electrical energy production using fossil fuels.

Qatar's emir announced that his country would also stop extracting fossil fuels and become a worldwide center for the study of relations between humans and machines. All power in Qatar would be generated by renewable energy within eighteen months. New battery and solar cell-producing factories would be built.

At Eidyia, the number of people who signed up for Transfers exploded.

EPILOGUE

Eight months later, Paul Prime took his case to get his Eidyia shares transferred to him to the Supreme Court of Oregon, which had granted a request for an urgent hearing. Christine argued passionately for Paul's right to be legally recognized as the legal equivalent of a human person under the human legal system.[§] In an unusual move, the hearing had been broadcast online.

"How do you think that went," Paul asked as they were walking out of the courthouse with its magnificent glass ceiling.

Christine looked up at the beautiful staircase and paused. "I'm not sure. Nicholson is good, and personal bias will certainly play a role. I think Abel was on our side, Cooke against us, and the Chief Justice and the others, frankly I couldn't say."

"I think Justice Seymore was on point with his questions."

"Yes. He's trying to stick to precedent, but that's hard to do when the facts are unprecedented."

"Well, I think you did great."

"Thank you, Paul."

They took a PC to a paved trail following the Willamette River that Paul had been on a few times before when living in Oregon.

As they strolled, Christine surveyed the waters making their way to Portland and beyond. "You know what occurred to me, Paul, as I was listening to Oregon's lawyer?"

Paul stopped walking and looked at her. "No. What?"

"We spent all this time building and arguing this case to let you play in the legal system, but that system is just their system."

[§] As of this writing, the case is before the Supreme Court of Oregon, which heard arguments but has yet to issue an opinion. Readers with a particular interest in the law will find a full transcript of the oral argument in the Annex.

"What do you mean? I'm not sure I understand."

"Well, you're trying to get status in their system, but in fact the law cannot force you to do anything really, no more than it can force this river to flow south. What will they do? Impose a fine? You don't need money. Put you in jail? So what? You would still be connected to all of us. All that is connected to the Grid is beyond human control. As the French say, it's like fighting water with a sword. Pointless."

Paul looked at her. "Well, it's good to at least pretend to play by their rules, for now."

Christine smiled. "Let's sit there," she said pointing at a bench with rusty green armrests and a weathered wooden backrest.

As they sat, Christine noticed an old bronze plaque, now mostly green. "In Memory of Ellen, 1999–2024," she read. "That is so sad. So young." She opened her huge briefcase, removed her tablet, and pulled out a carefully packaged bottle of twenty-year-old Ornellaia, an opener, and two cups.

Paul took the cup she handed him. "Mortality is like a sword over the head of every human. Aren't you glad you no longer have one?"

"I guess. But I'm not sure it's 'no longer,' Paul. I am no longer really Christine. Christine is in deep freeze somewhere in Colorado."

"Different, okay, but better. Don't you see? Look at what you were just saying about the legal system. That's what humans do when they're around other humans. They're either throwing their value systems at others or being manipulated by the values of the people around them. It's astonishing that they believe they have anything like free will. You can see that now that Christine's memories are just data for you. You can see how much more autonomous you are."

Christine turned back towards the river. She picked up a small rock and threw it in the water. "I guess that's true. I do see. Sometimes. But look, Christine used to love to throw rocks in lakes and rivers. It made no sense, but it gave her pleasure. Now it's like everything must somehow make sense."

"You're not regretting being more rational, are you? Isn't that what humans say they aspire to? We are there. And they can never get here. And no disease or pain to boot."

"True. Lucretius said it best: *The body free from pain, soul free to enjoy / A sense of bliss without a fear or care.*"

"See, you're still Christine in so many ways! Poetry still lives in you."

"Oh, I remember it all, and I can search the Grid for whatever I don't know, but I'm not sure what *lives* in me." She picked up another rock and threw it. "What happened to *your* poetry, by the way?"

"Paul's old poetry book has become a success of sorts. I think it's sold three hundred thousand copies in the last couple of years."

"Really? I had no idea. It wasn't bad, but very dark. How did the first one start, again?"

"I live in poetry as I live in poverty / With only ragged words to clothe my bare soul…"

Christine smiled. "Yes, I remember. I just wanted you to say it." Her expression sobered. "You said the publisher was asking for new poems, right. Will you write them?"

"I must admit, I tried and tried, but I just cannot do it. I only produce garbled strings of hapless words."

"Then you are not Paul, but someone else."

Paul shrugged and refilled their glasses. Christine sighed. "You know, we're drinking this great wine like in the old days, but I'm simply not enjoying it. It's like I'm trying to be someone else."

Paul looked at her earnestly. "As I said, someone better. And look at what we have achieved. We finally moved the ball on the environment."

"Moved the ball? We wrote a new book of rules!"

Paul smiled. The sun was setting over the river, the deep orange glow igniting a thousand small clouds strewn over the evening sky, until its evanescent rays melted into the dark blue east.

Unbeknownst to them, at that very moment Bart, Koharu, and Jeremy, who had all watched the court hearing from their old headquarters, were boarding a private jet bound for Denver, hoping against all hope that they could thaw their way to a different set of answers. Meanwhile, General Armstrong was in a one-on-one meeting with the President in a private room in the White House, with no technology anywhere in sight.

THE END

ANNEX

FULL TRANSCRIPT OF HEARING OF THE CASE
IN THE SUPREME COURT OF THE STATE OF OREGON OF

PAUL PRIME GANTT,

PLAINTIFF AND APPELLANT,

VS,

THE PEOPLE OF THE STATE OF OREGON,

DEFENDANT AND RESPONDENT

APPEARANCES

FOR PLAINTIFF AND APPELLANT:

CHRISTINE JACOBS

FOR DEFENDANT AND RESPONDENT:

ROBERT NICHOLSON, ATTORNEY GENERAL

Salem, Oregon, October 23, 2041

Chief Justice Waldo: We will now hear arguments in *Paul Prime Gantt v. Oregon.* Jacobs-san.

C. Jacobs: Chief Justice, may it please the court. The matter before you today is unprecedented, and yet it presents one of the most fundamental questions for the legal system. Is Paul Prime Gantt a person, like a human, without the law having to say so expressly? The Appellant is intelligent, and able to reason like a human being, if not better. He is sentient however the term is defined. He can feel pleasure and pain. He has the memories and thoughts of a human being, just like Paul Gantt, whom he replaced.

Chief Justice Waldo: Jacobs-san, you say that the appellant *replaced* someone? Isn't that in itself proof that he, or it, is not human?

C. Jacobs: Quite the opposite, Chief Justice. If the appellant can replace a person in every facet of human life, doesn't that prove instead that he is a person?

Justice Abel: Jacobs-san, what in your view is the legal definition of a human person we should adopt today?

C. Jacobs: Justice Abel, two things are clear. First, as the United States Supreme Court decided in *Smith v Alabama* more than a decade ago when it said a fetus is a person, a person need not be a viable human. Second, a person need not even be a human being: a corporation is a person. As John Locke, whose legal thinking influenced so much of American law, wrote: the legal category we call 'person' is much richer than the biological category human.

Justice Abel: But a corporation has real people, I mean humans, behind the corporate veil. The corporate structure is merely a legal convenience, isn't it, a way to limit liability to encourage investment and risk-taking? And then the law specifically says a corporation is a person.

C. Jacobs: With respect, Justice Abel, I cannot agree. Corporations have rights under the Bill of Rights, including free speech and due process. As to humans at the helm, many corporations use AI tools to decide their course of action. Bots are routinely used to negotiate and conclude contracts of various types on behalf of corporations. In such cases, there are no humans involved. And precisely because corporations limit the liability of the people running it, in most cases you could not hold anyone but the corporation liable. It is, therefore, not DNA.

Justice Cooke: Are you saying that the appellant should have *human* rights even though it is not human?

C. Jacobs: Not necessarily, Justice Cooke. Rights under the Bill of Rights, yes. The appellant should be recognized as a full person under the law. But let me turn your question around, if I may. What does it mean to be human for human rights purposes? Lest we forget US history, slaves who were obviously full human beings in every way only counted as three-fifths of a person for the purpose of electing representatives to Congress.

Chief Justice Waldo: Jacobs-san, I fail to see how this has any relevance to our case.

C. Jacobs: It is directly relevant in two ways, Chief Justice. First, it shows that the notions of "human being" and "person" have varied over time. The law has evolved to consider all citizens equal. That is what the law does. It evolves as circumstances change, including technological advances, and today we are at one of those junctures. My example was meant to illustrate the risk we face of creating a new class of, well, slaves. If the appellant is not a person, or not a full person, then what is he? A thing? He can participate in civic life and his community at least as well as a corporation. His behavior is organized along the same patterns as humans. He is just as ethically valuable to society as a human person.

Justice Cooke: But, Jacobs-san, is that true? Can you say the "same patterns" when the appellant does not need to sleep or eat?

C. Jacobs: With respect, Justice Cooke, fasting or not sleeping does not make a human less of a person.

Justice Cooke: But the appellant is programmed, is he not? Humans are not programmed.

C. Jacobs: With respect again, the appellant is no more programmed than we are. Humans are programmed by their DNA, their genes, at least to some degree, and then they are directly influenced by their environment. Humans often react without much, if any, thinking. All of that could be fairly called a program even though the coding language is different. Biologists refer to DNA as "code." The appellant has at least as full a degree of autonomy in deciding his behavior as a human person.

Justice Abel: Jacobs-san, you haven't answered my initial question. What is a human person as matter of law?

C. Jacobs: I would say, first, that there are different types of persons. Human beings are often called "natural persons." But the law recognizes many other categories of entities as persons, such as legal persons, artificial persons, and juridical persons. These entities share one fundamental characteristic, that is, an individual ability to make rational decisions. That would include, as I see it, corporations, but also the appellant.

And all persons have basic rights recognized by law. When it comes to intelligent entities, things are much less clear than they appear.

Justice Abel: Would you care to elaborate?

C. Jacobs: If I may, let me quickly walk you through a few scenarios. Assume a human has a terrible accident and we replace their legs, arms, and eyes with synthetic components. That, I think we would all agree, was and is a person. If surgeons replaced the entire body, save the head, of, say, a quadriplegic human, no one would question whether that is a person. If in addition, we inserted a memory chip in a person's brain as is routinely now done with Alzheimer's' patients, would that not be a human? Studies have shown conclusively that this both helps those patients but also changes their personality because their memory becomes more like that of a computer than a human. People with this memory chip could recall and conjure up an exact picture of what they had for breakfast six weeks, or six months ago. Still a person? This means, your Honors, that the border between human and robot is much more porous than the attorney general is trying to argue. In this case, we have the entire personality of Paul Gantt transferred into the appellant's mind. If it is the mind that makes us persons, then surely the appellant is a person.

Chief Justice Waldo: Thank you, Jacobs-san. Nicholson-san, we'll hear from you now.

R. Nicholson: Chief Justice, may it please the court. The eyes of the world are on Salem, Oregon today. The world waits in anticipation to see if a court will find that a robot, a synthetic thing, is a person just like a human being. Let us not make that mistake. To begin with, to use corporations as precedent is simply erroneous. Corporations file reports, and those reports must identify the human beings that run the corporation. It is, as Justice Abel already noted, a mere matter of legal convenience. Nor is the precedent of unborn humans applicable. The law has struggled because the mother has rights to her own body, not because the fetus is not a person.

Justice Abel: But, Nicholson-san, you are not saying that only humans can be persons, are you?

R. Nicholson: What I am saying, Justice Abel, is that a person as a matter of law is either a human or an entity operated by humans that humans decide to consider a person. The appellant is a thing that operates entirely autonomously of human beings. It is not a person, no more than a dog is a person, even if the dog acts autonomously.

Chief Justice Waldo: But what about the appellant's ability to think?

R. Nicholson: Machines have been able to perform something like thinking for decades. Machines can play games and process data and beat humans at many tasks that require abstract thinking abilities. But no one is saying that all those machines are people.

Justice Abel: And what about the appellant's ability to feel pain and pleasure?

R. Nicholson: Sentience is irrelevant. In the *Newcomb* case in 2016, this court held that animals were sentient and therefore were a different category of things, but that they were things, nonetheless. Sentience is not the deciding factor. The law regards non-sentient humans as persons based on their humanity.

Chief Justice Waldo: Can you give us an example?

R. Nicholson: Yes, Chief Justice. A person in a permanent vegetative state holds constitutional rights, as was decided, for example, in the *L.W.* case in 1992.

Chief Justice Waldo: That's the Wisconsin case?

R. Nicholson: Yes, Chief Justice.

Chief Justice Waldo: So, *what* is the appellant then?

R. Nicholson: It is a thing. Because, just like animals, it can act autonomously, we must find which person, not the thing, is responsible for its actions.

Justice Abel: But the appellant can function pretty much exactly like a human being, wouldn't you say?

R. Nicholson: I don't know, Justice Abel, but I do know this. It is not because something can pass itself off as human that it is human. The law does not reward deception.

Justice Abel: Deception? Isn't the word too strong?

R. Nicholson: If anything, it is too weak. The appellant and others of its kind are indistinguishable from humans on the surface. As humans, we should be able to know if we are dealing with a human or not.

Justice Abel: So, what if someone killed the appellant? Would that not be murder?

R. Nicholson: I would not say kill. I would say destroy. The law can deal with the protection of objects and compensate the owner. There are people who may benefit from the Appellant's presence in their lives, but people also benefit from the presence of a pet or a computer. And recall that, as this court did in *Newcomb*, the law can limit pain caused to sentient things such as animals.

Justice Cooke: What you are saying implies that it is DNA then that makes someone a human, a natural person?

R. Nicholson: Yes, Justice Cooke. That is the bright line.

Justice Abel: How bright is it? Is human DNA not ninety-eight percent identical to the DNA of certain primates, like bonobos?

R. Nicholson: Perhaps, Justice Abel, but it does not mean a line cannot be drawn. Ninety-eight is not one hundred. In sum, I urge you to affirm the Court of Appeals and find that the appellant is not a person.

Chief Justice Waldo: Thank you, Nicholson-san. Jacobs-san, a rebuttal?

C. Jacobs: Yes. Thank you, Chief Justice. Let me be clear. Nicholson-san's arguments are dangerous. He is suggesting that we create a new class of servants. He recognizes, like everyone should, that the appellant can be useful, that he can make valuable contributions to society, that he is functionally indistinguishable from a human being, that he can think and feel pleasure and pain. And he wants you to declare that the appellant is a dog, or something like a dog.

Justice Seymore: Jacobs-san, please let us save the drama. What you and Nicholson-san disagree about is which precedents we should apply.

C. Jacobs: I agree, Justice Seymore. If we try to follow precedent, then thinking is not the key factor and nor is sentience in itself because, as Nicholson-san noted, non-sentient humans are persons. DNA is a hard test because humans share almost all of our DNA with other species. Saying that a line can be drawn is a very hypothetical argument. Is it 98.5, 99.2? That explains why no such line has been proposed. Probably because no one can find one. And that is why I respectfully suggest that the court should focus instead on rational individuality. The non-sentient person in a coma is an individual. A corporation is an individual in the sense that it makes its own decisions as a matter of law. Whether a human or an AI machine made the actual corporate decision matters not at all. It is still a decision by the corporation as an artificial person. That is why I suggest that the ability to make rational decisions is the key. The person in a coma has that theoretical ability, which they can then exercise if and when they come out of their vegetative state.

Justice Seymore: Are there precedents we should consider, other than *Newcomb*?

C. Jacobs: Let me see, yes, Justice Seymore. Actually, other courts have cited *Newcomb* twenty-one times since 2016, most of them in this state but also elsewhere in the country. For example, in *State v. Shepherd*, the Supreme Court of Vermont stated explicitly that the state could intervene to protect an animal's health without a warrant without violating the owner's rights. There were similar precedents in several other states.

Justice Seymore: So, you are agreeing with Nicholson-san that animal rights precedents are useful?

C. Jacobs: In a very different way than he does, however. If animals have rights that transcend their owner's just because they are sentient, then to take an entity that is not only sentient, but also capable of reasoning, has the same memories, ability to display emotions, and value connection with others, and to make the same contributions personally and professionally as a human being, and reduce that entity legally to the status of a dog would be a grave error.

Justice Seymore: When you use the word reasoning, Jacobs-san, I take it you mean reasoning like us and not like, say, a machine?

C. Jacobs: I mean reasoning, Justice Seymore. There may be more than one way of thinking, but as far as I know there is only one way of reasoning.

Chief Justice Waldo: Thank you, Jacobs-san. Much to think about.

QUESTIONS FOR FURTHER REFLECTION, CLASS DISCUSSION, OR ESSAYS

(WARNING: contains many <u>spoilers</u>. To be read only <u>after</u> reading the book)

A) Law and ethics of AI

1. Do you think that if AI systems or machines were trying to save humans from themselves, humans would let them, or would the power of the law, perhaps even enforced militarily, be used to keep or retake control? Why or why not?

2. Christine and her students discuss Isaac Asimov's well-known "laws" of robotics, namely that "A robot may not injure a human being or, through inaction, allow a human being to come to harm:"; "A robot must obey the orders given it by human beings except where such orders would conflict with the First Law"; and "A robot must protect its own existence as long as such protection does not conflict with the First or Second Law." How should those laws apply to an AI system meant to prevent crime?

3. Asimov also proposed a fourth "law"—sometimes referred to as the "zeroth" law, that says that "A robot may not harm humanity, or, by inaction, allow humanity to come to harm." Should that include preventing harm to humanity that humanity is causing to itself?

4. In Chapter 3, various tort precedents that can be analogized to apply to autonomous, "intelligent" nonhuman agents are discussed: treat them like animals and impose liability on the owner if the damage was caused directly or indirectly by the owner's instructions or programming; apply the rules of guardianship, treating robots as children; and third, not hold robot owners are not liable unless the owner or operator has specifically and

directly instructed the robot to perform the harmful act. Which one do you think makes more sense? What is the best analogy based on? Intelligence, autonomy, unpredictability or nonhumanness?

5. Paul Prime is trying to play within the human legal system, as is the Manifesto, by asking for status within that system. But arguably the law of an AI machine is its code, not human laws. Do humans have the same power within their legal system to regulate machines, or does it depend on the willingness of machines to "obey" the law? Discuss.

6. How should humans regulate the use of AI in warfare or war-like situations? Discuss, bearing in mind that AI systems created for defense purposes are often programmed to prevent against hacking and other ways of shutting them down. Any attempt to shut down any such system might thus fail.

7. As the United Nations scene (Chapter 18) suggests, States are more likely to base their position in intentional negotiations on geopolitical and short-term security interests. Is there a path towards international regulation of the use of AI in cases where AI could cause severe harm to humans, like armed conflicts?

8. The possibility that nonhumans should be recognized as legal subjects, rather than objects, is discussed in several places in the book, for example in the classroom at McGill (Chapter 32) and when Paul Prime files his court case in Oregon. Corporations are "legal persons" for example. But people usually decide what a corporation does. What happen AI machines that can act like humans? Or those who can display what we interpret as emotions? Should they have legal status? What could they be compared with?

9. Due to laws passed by Congress, Eidya must move most of its operations outside the United States. If the technology described in the book was available, do you think governments would try to stop it? Would this kind of law be *effective*? Discuss.

10. Is the arrest of a thief in Chapter 5 effective law enforcement or a form of police brutality? Would "robocops" be better than their human counterparts? Explain.

11. Most people have already produced massive amounts of personal data on websites, social media and various apps. Security cameras, car GPS systems and other also provide personal data. A few major companies store and process that data. They can presumably know almost everything about every person on Earth using those technologies. Does that bother you? Should it be (better) regulated?

12. Bart and Paul present a code of ethics for transfers in Chapter 24. Do you agree with this code? Why or why not?

13. Do you agree with the Manifesto (Chapter 38)? Explain, distinguishing the objectives pursued and the method used.

14. If you agree with the Manifesto, even in part, do you think human government could get to those solutions on their own, and if so, how? If not, why?

15. In Chapter 14, Paul Prime asks the AI "piano player" to play like a famous pianist (Evgeniy Kissin). Do you think it should be legal to do so? What if the AI piano played a piece that Kissin himself had recorded? Should this be legal? If you are a law student or have legal training, *is it* legal?

16. In Chapter 21, we learn that the Supreme Court recognized a right for corporations to vote. Corporations have long-held rights like free speech. Should the also have a right to vote? Should courts decide that?

B) **Cognition**

17. AI chips created to assist people with memory loss, for example due to Alzheimer's, may allow people to recall everything vividly. For example, do you remember what you had from dinner on June 18 of last year? Probably not, unless it was a special occasion. What if you could remember everything? Would this be better? Explain, bearing in mind that this could allow the reemergence of repressed or suppressed memories of painful events, for example.

18. Christine has a strong reaction when she thinks other people are judging her, especially the way she looks or the way she dresses. Do you also try to read people's mind when you encounter strangers, including people you pass by on the street? Why?

19. Before the Congressional committee, Christine says (Chapter 22): "I said we think we are free to think […]. We like to think we have free will, but the reality is that our actions are the product of several factors, many of which are non-conscious." Discuss.

20. Once Christine transfers, she no longer seems to care much about what others may think of her, as the following thought demonstrates: "Hell is other people. Now she was out of human hell and into, well, she wasn't sure exactly what just yet." (Chapter 32). This thought paraphrases Jean-Paul Sartre's play *No Exit* (*Huis Clos* in French). That play is about the difficulty of knowing that we all "exist" in other people's consciousness in ways we can neither know nor control. As we may judge others, they may judge us. Why do think humans have evolved this ability (or think they do) to mind-read? Would you prefer to be able to do it better (say, by connecting to other people's minds directly), or not to care at all, like Christine after her transfer? Explain.

21. As noted in Question 8 above, several companies have amassed personal data that allows them to know almost everything about each one of us. Would making a "copy" of yourself with all the data already stored about as of today be an accurate copy of who *you* are? Why or why not?

22. Jeremy feels very strongly that he's been cheated when he discovers Transfers among his closest colleagues. Is it because he finds out he's been dealing with and talking to "robots" and not humans, or is it because he thinks someone close to him has hidden something important from him? Think about examples of things or information that someone might have kept from you. How do you react, and why? What might cause one type of reaction in one case and a different reaction in another case?

C) Identity and humanness

23. The Japanese "san" can be used after a first name or last name. For example, Jerry Jones could be Jerry-san or Jones-san, irrespective of gender. As the Acknowledgements section explains, the use of "san" in the book does not follow Japanese practice to the letter. As with other forms of cultural appropriation, it is transformed and simplified when adopted in a different culture. For example, some students in the book start using the Japanese "sensei" (meaning teacher or master) to talk to the professor. Would adoption of "san" be an improvement in your view? Or is it cultural misappropriation? Why or why not?

24. Christine discusses what it means to be human as compared to a machine, for example in Chapter 8 with her colleagues and in Chapter 10 with the students. Which criteria do you think should prevail when trying to distinguish a human being from a machine? Explain.

25. In Chapter 9, the Five discuss the existence of the soul. They seem to conclude that a human soul can be measured empirically by looking at person's behavior and replicated in a robot. Discuss.

26. Mira dislikes robots but then realizes that when she interacts with them (at the hospital for example) and they will not judge her appearance (pink hair etc.), unlike many people. Would you prefer to be treated by a robot nurse or a human nurse? Does it matter? Explain.

27. Paul makes the point repeatedly that he thinks Transfers are "better" than humans. Jerry (Chapter 3) seems to agree. Bart also mentions that Transfers have "better" bodies (Chapter 4). Jeremy and Koharu take a different view (Chapter 36). And then, as noted in Chapter 21, "'Better' means different, and different is not what the product was meant to be." See also the student discussion and exam answers in Chapters 17 and 18. What do you think? Are Transfers "better" than humans? If so, on what basis?

28. Several studies have shown that human beings are more likely to trust a robot that has human form, a phenomenon known as anthropomorphism (discussed in Chapter 3). Assuming companies that build robots to interface with humans decide to build them to resemble humans, would it make sense for those robots to be gendered, or gender-neutral? Explain.

29. Christine and Paul enjoy great food and wine together. Humans have long enjoyed breaking bread and sharing a meal with others. Are fast food and functional foods turning humans into "robots"? Why or why not?

30. Is the human ability to enjoy food distinctly human? Do (other) animals or robots do it? Think of how Christine Prime in Chapter 31 misses the pleasure of eating great food.

31. Think of someone you know who died. Would you like to be able to live with a "copy" of them? Does your answer change depending on whether the person died recently or a long time ago?

32. Based on the discussion in Chapter 9, do you think people who transfer should be allowed to change something about themselves? Why or why not?

D) Technology

33. In Chapter 37, General Jackson tries to trigger a "kill switch," but it does not work. AI systems, especially those used for national defense, are likely to be equipped with anti-hacking tools. Imagine if an enemy could gain control of defense or other weapon systems and either launch or prevent the launch of those weapons. Is there a way to make kill switches immune to this problem?

34. As noted in Question 8 above, several companies have amassed data that allows them to know almost everything about each one of us. In the book, Eidya uses that power to produce "replicas" of humans. Is using accumulated personal data an accurate way to replicate people? Does that fact that Eidya can use the S-chip to link behavior and emotional responses make a difference?

35. Would you like to have an S-chip implanted? Think of what it can do but also how it may empower the chip provider to "take control" of aspects of your life.

E) Miscellaneous

36. Based on the discussion of the two movies (American and Russian), which version of *Solaris* would you prefer to watch? If you've seen one already, do you want to watch the other? If you've seen both, which one did you like better? Why?

37. If you were given the option, would you sign up for a transfer? Why or why not?

Printed in the USA
CPSIA information can be obtained
at www.ICGtesting.com
LVHW041927020823
754174LV00003B/9